TAN YOUR HIDE!

TAN YOUR HIDE!

Home Tanning Leathers & Furs

by Phyllis Hobson

With a Special Section on

Working With Leather

by Steven Edwards

Storey Publishing

The mission of Storey Publishing is to serve our customers by publishing practical information that encourages personal independence in harmony with the environment.

Illustrated by Steven Edwards
Designed by David Robinson

Printed in the United States by Versa Press
50 49 48 47

Library of Congress Catalog-in-Publication Data

Hobson, Phyllis.
Tan your hide!
ISBN 978-0-88266-101-8
Includes index.
1. Tanning. 2. Leather work. I. Edwards, Steven M. II. Title.
TS967.H58 675 77-2593

CONTENTS

PART I

HOME TANNING LEATHERS & FURS

PART II

WORKING WITH LEATHER

PART I

HOME TANNING LEATHERS & FURS

by Phyllis Hobson

HOME TANNING: PRO & CON

In these plastic and vinyl times, few materials are more luxurious than a piece of well-tanned genuine leather or a soft, warm fur.

But the price of these materials becomes more and more out of reach for the home craftsman who likes to work with leather or the seamstress who likes to sew with fur. By the side (half a hide), leather costs more than $4 per square foot and even tanned rabbit skins are selling for $3 and $4 each.

For the homesteader who raises the meat for his table, or the hunter who shops for his meat in the back woods instead of in the supermarket, the answer may well be to tan the animal hides which otherwise would be wasted and from them make the leather and fur items for his own use.

Tanning hides at home is not an easy job. There is a great deal of manual labor involved in skinning an animal, then working with the hide to produce a fur or a piece of leather.

But if you enjoy the luxury of working with a good piece of leather or a fine fur, home tanning can be a source of great satisfaction and pride. It is good to know you have mastered a new skill. It is satisfying to know you have created a beautiful, useful piece of leather or fur from a hide that otherwise might have been discarded.

The art of tanning animal hides into furs and leather dates back to the beginnings of civilization.

In the Old Testament, Abraham is described as giving Hagar a water bottle of leather, which probably was tanned with bark. In *Genesis,* it is said, "Unto Adam and also his wife did the Lord God make coats of skin and clothed them."

Leather was mentioned in Egyptian records nearly 5,000 years ago, and specimens (tanned with alum) which are at least 3,000 years old have been found in China.

An early tanning recipe recorded by the Arabians instructed the tanner to cover the skins with milled wheat and salt for three days, then scrape off all pieces of fat and flesh from the inside. The stalks of the Chulga plant were then pounded to a pulp and dissolved in water. This was applied to the outer side of the skin and left for twenty-four hours until the hair fell off. The skin was then allowed to set two or three days to complete the process.

Another home tanning recipe which dates back to the 1880's calls for soaking the hide in the diluted solution of slaked quicklime to loosen the root hairs, scraping off the hairs and then immersing the hides in a solution of dung, which contains the pepsin juices to partially digest the soft substances on the hide which must be removed. Nothing is said in the recipe about methods of removing the odor of the dung.

One very old method of tanning was used by the Crow and Navajo Indians, who taught the first settlers of this country their method of "buckskin tanning."

The Indians loosened the hair by rubbing the skin with a lye paste made by adding water to the wood ashes from their campfire. After the hair was scraped off, the skins were rubbed with a mixture of raw brains and liver from the animal that had been skinned.

Indian fire for smoking buckskin

The flesh side of the hide was then scraped with a thigh bone taken from the animal and used as a fleshing tool. After the skin had been rubbed and worked until soft, it was smoked over a low, smoldering fire to preserve it and give it a pleasant odor.

There are two recipes for making buckskin — by the old Indian method and by a new adaptation — on pages 64-65.

Modern tanning methods make use of chemicals instead of vegetable tannin or animal fat to preserve hides. Some of the chemicals used in tanning, such as lye, lime and acids, can cause serious burns or even blindness.

For reasons of safety and accuracy then, there are three rules which should never be broken when tanning hides with chemicals:

1. Weigh or carefully measure all ingredients.
2. Always wear gloves when handling hides or chemicals.
3. Treat all chemicals with respect.

Note also (page 57) the dangers of tanning solutions.

WARNING!

Some animal diseases, such as tuberculosis, tetanus, rabies and anthrax can be transmitted through the handling of infected animals. Any animal found sick or dead or acting strangely should not be skinned, but should be buried or otherwise disposed of to keep it from contaminating people or other animals. Do not handle it with bare hands. Do not skin it, even with gloves on.

Uses

The hides of certain animals generally are suitable for particular uses. The hides of young calves and goats, for instance, are thin and supple enough for women's dress gloves, but would be too thin for shoe leather. The much thicker hides of mature cattle and older horses are heavy enough for soles and belts but are not suitable for gloves or clothing.

The species as well as the size and age of an animal will determine the thickness of the hide and therefore its best use. Skins from small animals such as rabbits, squirrels and very young sheep or goats are usually too thin to be used for leather, and are best used for furs for small items or pieced together to make coats and jackets.

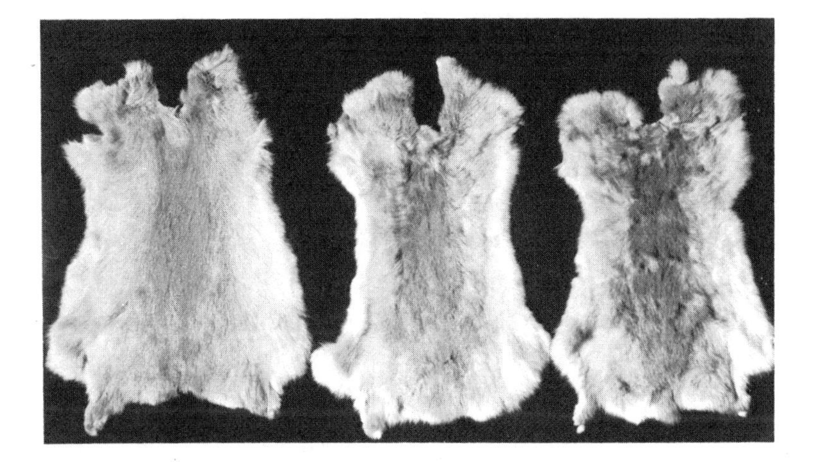

Rabbit pelts, beautifully tanned and soft, ready for making into hats or mittens. (USDA photograph)

Medium-sized skins, called "kips," from such animals as calves, goats, full-grown sheep and deer, are heavier and may be used for light leathers suitable for making gloves, moccasins, handbags and shoe laces, as well as for furs.

Heavy skins, termed "hides," from large animals are suitable for shoe leather, belts and harnesses. Because of their weight, these hides seldom are tanned for furs except for lap robes and rugs.

What You'll Need

In spite of the length of the following list, home tanning can be accomplished with a minimum of equipment at low cost. The trick is to improvise the needed tools in your workshop or to salvage them from old barns and junk shops.

The principal investments are time — it takes two to six weeks to tan a hide at home — and muscle power. The softening process is accomplished by fleshing, scraping and rubbing the hide by hand, back and forth, over a tanning stake. And if you think that sounds like a lot of work, you're right.

The following is a list of the needed, (then the desirable) materials and equipment:

Animal Hides. You'll need a supply of animal skins, wild or domestic, large or small, to tan. Since it's likely that you'll have some failures before you master the art of tanning, it's a good idea not to start out on a prize pelt or with a valuable animal hide. The best advice is to start small and tan the skins for fur, then work up to the larger, more valuable hides and learn to make leather. Domestic or wild rabbit pelts are ideal to learn on.

Containers. Tubs or barrels in which to soak the skins in the de-hairing and tanning solutions should be wooden,

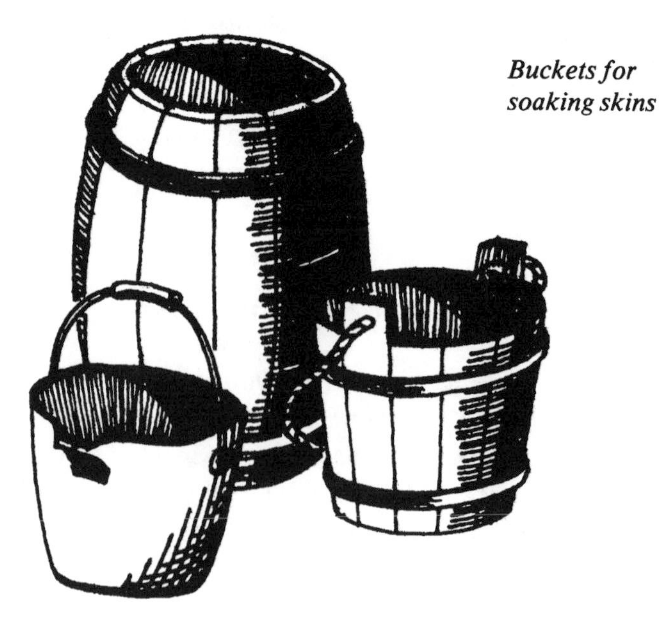

Buckets for soaking skins

earthenware or enamel. Do not use iron, galvanized steel or aluminum containers. The size of the container needed will depend on the size of the skins. Allow enough room for the skin to hang without crowding, but too large a container is a waste of expensive tanning solution.

Fleshing Knife. A tanner's fleshing knife is designed for the job of scraping the hair, fat and excess tissue off the hide without damaging it. The 15- to 17-inch blade is slightly curved, with a handle on each end. One side of the blade has a dull edge to scrape off the hair after liming. The other side has a sharp edge for the delicate job of shaving off flesh and tissue and thinning down the hide. A list of companies which manufacture tanning knives is on pages 75-76.

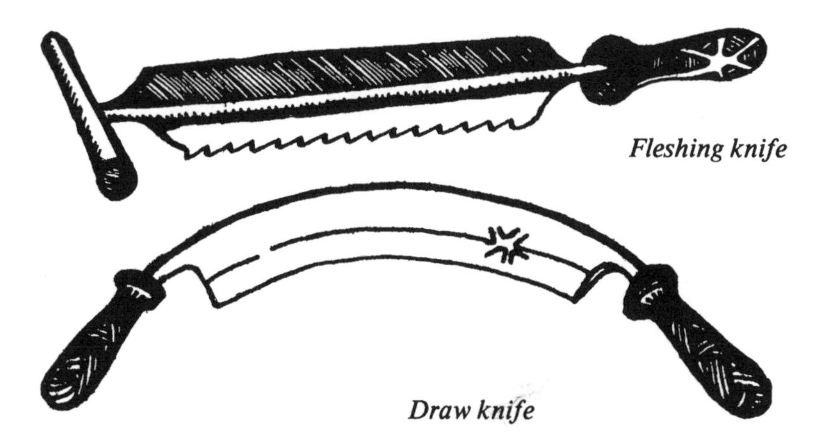

Fleshing knife

Draw knife

Draw Knife. A draw knife, a smaller version of the fleshing knife, has a curved, 8-inch blade with a bent handle on each end. This tool is useful in places where the fleshing knife might be too large, or for tanners who can handle the smaller knife more easily. A carpenter's drawshave is very similar and could be used in place of the draw knife.

Butcher Knife. Although it is not as convenient as the fleshing or draw knives, a 10- to 12-inch butcher knife may be substituted for a special fleshing tool. To do so, drive the pointed end of the knife as far as possible into a wooden handle, to make a handle on each end. The butcher knife also has a dull and a sharp edge.

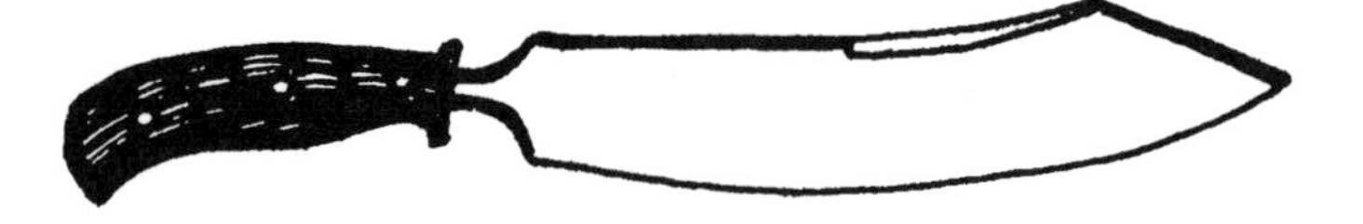

Butcher knife

Skinning Knife. A slender, razor-sharp skinning knife is important for the step which determines more than any other step the condition of the finished product — skinning the animal. It is possible to substitute a good butcher knife for this tool, but it will require much more care and skill to obtain a skin in good condition.

Skinning knife

Slicker. A slicker is a 5- or 6-inch wedge of metal or wood used to work out the leather after tanning. A suitable slicker may be hand-made by shaping a piece of hardwood 6 × 4 × 1½ inches and tapering one end to a dull edge.

Stake. A stake may be made of a 1 × 6-inch or a 1 × 8-inch board which has been thinned to a wedge shape with rounded

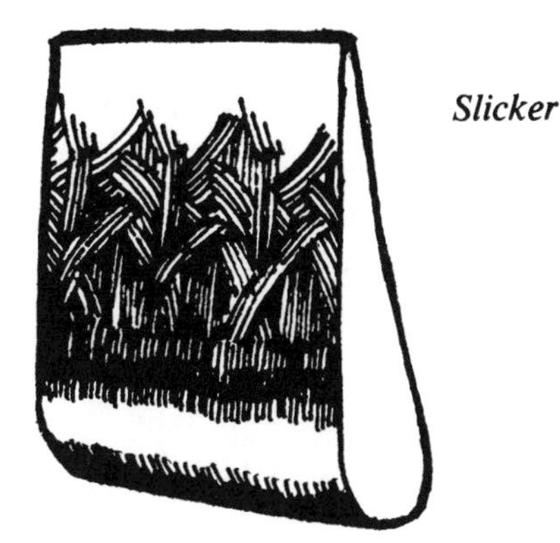

Slicker

corners on one end. It should be sanded smooth so the hide will not be torn or scratched during the softening operation. The stake may be three feet long or more, whatever is convenient for a standing operation. Or for sit-down work a stake 16- to 18-inches high may be attached to angle iron brackets and fastened to a table or bench. Or for simplicity, the stake may be pounded into the ground far enough to make it steady.

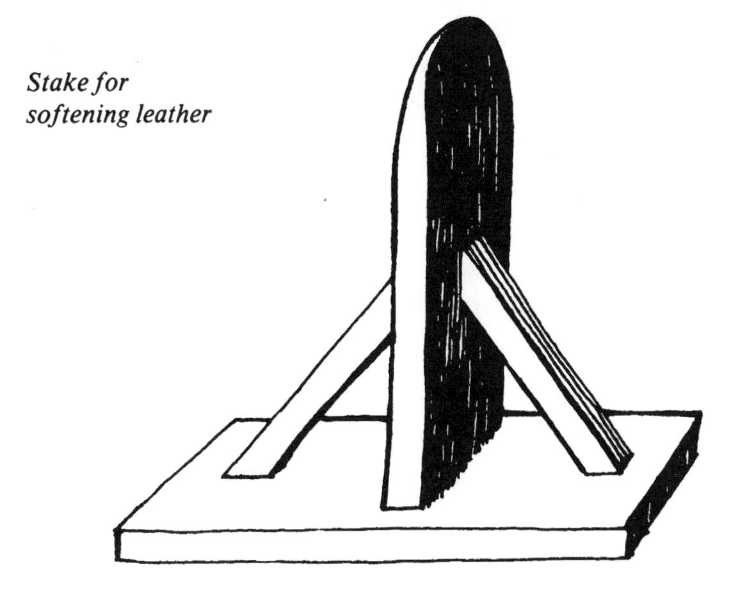

Stake for softening leather

However it is done, it is important that the stake be firm and steady in an upright position, for it must undergo a great deal of back and forth pressure. The longer stake may be braced with a board on each side and fastened to a firm wooden footing on which the operator stands, or it may be attached to the floor with wood or metal braces.

Beam. A tanner's beam is a rounded, slanted working surface on which the hide is spread for fleshing. For a large hide, an 8-foot-long log at least 2 feet in diameter is satisfactory. Debark the log, smooth the surface and raise one end up to working height. For smaller hides, a 2 × 12-inch board, rounded at the edges to simulate a half-log, is suitable. If necessary, a flat surface — a workbench or a table — may be used.

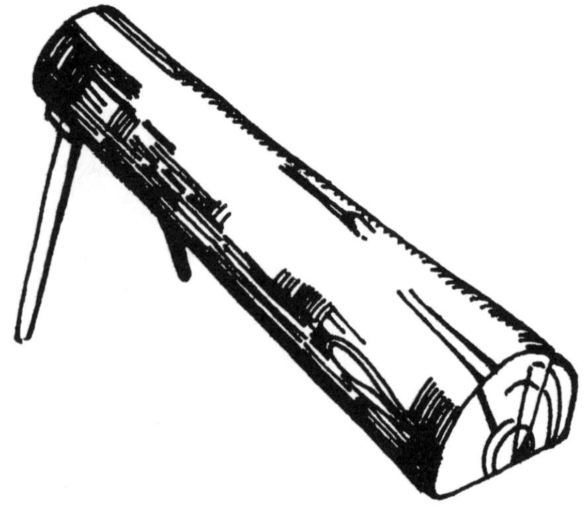

Tanner's beam

Rubber Gloves. A pair of heavy, elbow-length, heavy-duty rubber gloves. All of the de-hairing liquids and most of the tanning solutions are caustic, and you should wear the gloves while mixing or disposing of them.

Rubber gloves

Sanding Block. A block of wood, 6 inches square, on which a piece of sandpaper is tacked. This is needed to smooth the leather surface after tanning.

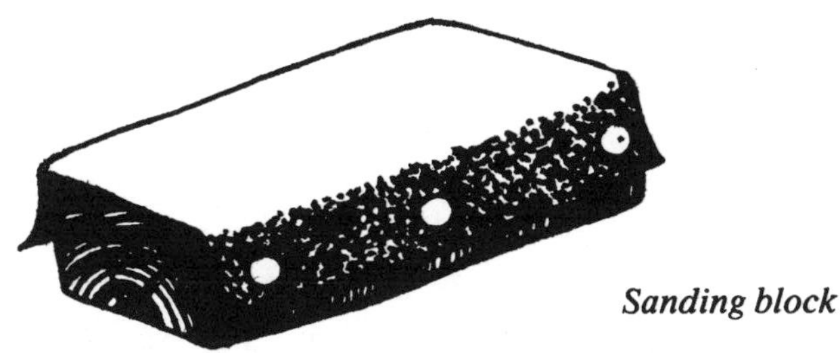

Sanding block

Tanning Chemicals. Many of the chemicals used in tanning and de-hairing processes are readily available around the home or at most drug or grocery stores. Salt used for salting the hides should be non-iodized table salt or the coarser pickling salt, which is easily obtained. Other chemicals, such as sodium bicarbonate and borax, are familiar household items. Some chemicals, such as vegetable tannin concentrate and chrome alum, may be ordered from chemical suppliers. A list of sources for tanning chemicals, materials and tools is given on pages 74-76.

Comb or Brush. Almost any good comb or brush will do. They are used for combing furs.

Plenty of Water. A great deal of water is needed for soaking, washing and rinsing the hides, as well as the water used in tanning solutions. A source of running water is a necessity, and one of the simplest is a hose connected to an outside faucet.

Access to Scales. Since the amount of chemical used is determined by the weight of the hide being tanned, the hide must be weighed before soaking. It is not necessary to buy this piece of equipment, however, if there are scales which can be borrowed occasionally.

Storage and Working Space. An even temperature is important through all stages of the tanning process, including later storage. It is necessary, therefore, to have a working and storage area that has good lighting, where spilled water is not

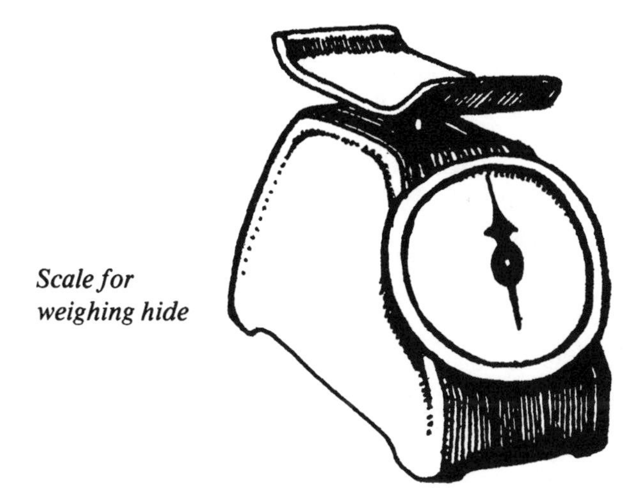

Scale for weighing hide

objectionable and with a cool, even temperature year around. A concrete-floored basement with drain and a water supply is ideal.

NOT NECESSARY, BUT NICE TO HAVE:

The following are not absolutely necessary, but if you plan to do a great deal of tanning, they might be good investments:

A Wringer Washing Machine. An older-type washing machine, the kind with an enamel tub and a center agitator with a wringer at the top (which will squeeze most of the water out of materials), is excellent for soaking, rinsing and wringing skins. It can do the job in half the time with half the effort. Used washing machines usually can be bought at low cost, but the leather-making process is far too messy to use the family washing machine. A discarded automatic washer, without wringer, could be used for soaking and rinsing.

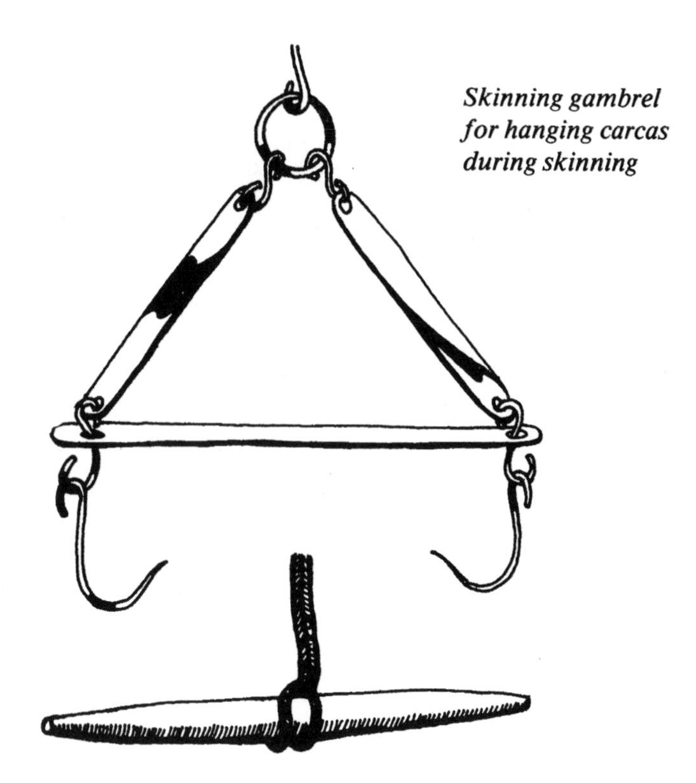

Skinning gambrel for hanging carcas during skinning

A Skinning Gambrel. It's possible, of course, to hang an animal on almost anything for skinning, but a skinning gambrel, made specifically for this purpose, makes the job easier.

A Clothes Dryer. A used home laundry dryer, one in which the heating element is burned out, sometimes may be available at appliance stores at no cost. If the tumbling drum is still operable, this piece of equipment can save you a great deal of hand work. The tumbling action which tosses the tanned skins around in the drying drum does an excellent job of softening them, much more evenly than you can do by hand and with a lot less effort.

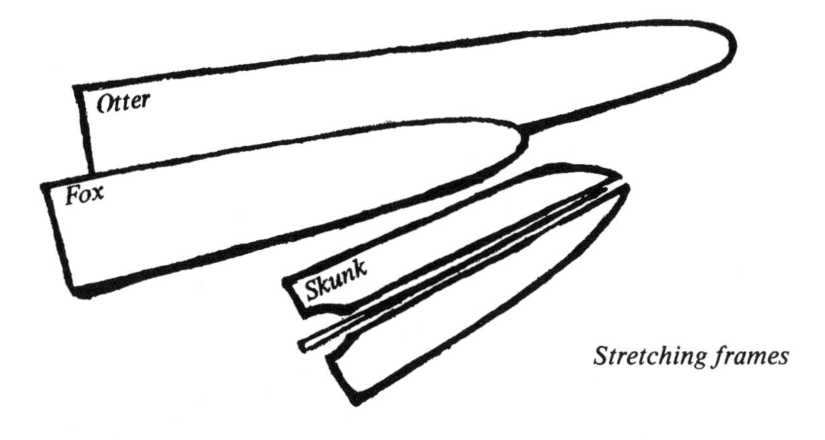

Stretching frames

Stretching Frames. Stretching frames are convenient for the many times it is necessary, after washing and soaking, to pull the skin back to shape. Stretchers may be made cheaply and simply, or you may stretch the skins without a frame by following the instructions on pages 21-23.

In addition to these equipment items, you'll need some patience and not a little muscle power. There can be a great deal of satisfaction in the result of home tanning, but the beginner should realize that there will be failures during the learning process. Home tanning of raw hides is a process best learned by practice. There is pride in a job well done, but there is no magic solution for instant tanning without work.

"Tanning" or "Tawing"?

Strictly speaking, the correct word is "tawing" when hides are preserved with alum. The term "tanning" originally applied only to skins which were treated with an infusion of tannic acid, a derivative of oak or other tree bark or of nutgalls.

"Tawing" was the proper word for the process of working skins after they had been soaked in a solution of alum and salt. But, except for the oldtimers in commercial tanneries, the term "tawing" is seldom used today, and "tanning" is commonly accepted as the process of preserving hides — by whatever method.

So throughout this book we will use the term "tanning," even when we know that the old tanners' term was "tawing."

Leather or Fur?

The purpose of tanning is to soften and preserve the hides of animals in order to render them permanent and useful, either as fur or as leather.

The methods of tanning furs and leather are much the same except for the de-hairing process which must be done before hide can be converted into leather.

To tan a hide for fur, it first must be cleaned and softened, then treated to preserve it. Natural tannin materials (vegetable tannin) are not used to tan furs today because they stain the fur. Alum often is used because it does not change the color of the fur and it acts quickly, thus avoiding loosening the fur. In addition, alum has the advantage of shrinking the skin, which tightens the follicles and better retains the hair.

To convert a hide to leather, it first is necessary to loosen the root hairs so they may be scraped off without damaging the animal skin. This de-hairing step is done after the hide is salted. All four tanning methods — vegetable, mineral, oil or a combination — are used for leather according to the thickness, the intended use and the tanner's preference.

Selecting the Hide

The exact weight of a hide will depend on the size, age and condition of the animal. However, the following are approximate weights of the hides of the following animals:

Mature steers, cows, horses	60 pounds
Two-year-old calves, horses	40 pounds
Young (6-8 months) calves, horses	8-15 pounds
Mature sheep (sheared) or goats	8-10 pounds

Animal with lines for "cut-off" and resulting hide

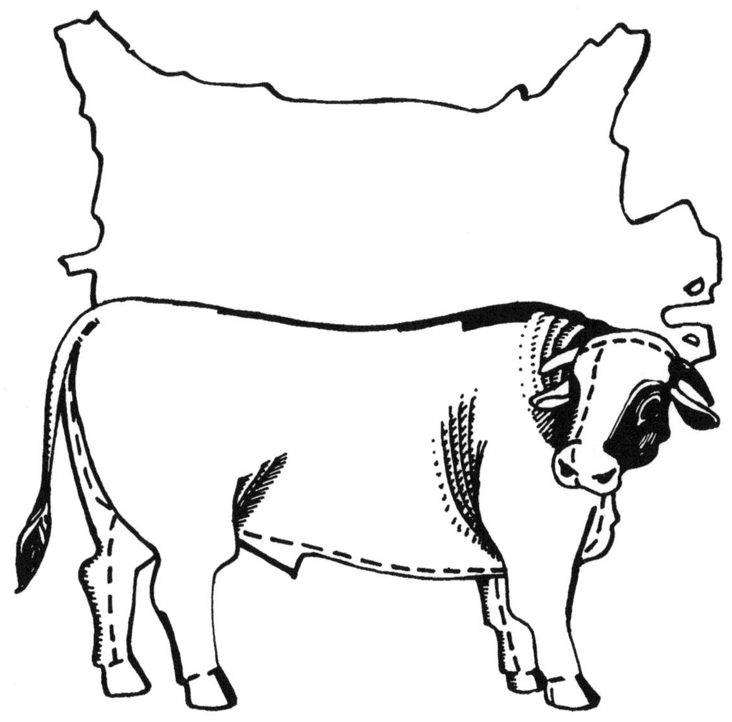

Trapper Ivan Anderson shows the variety of furs — otter, muskrat, raccoon — that he traps in Nebraska. (SCS photo — Herb Kollmorgen)

To obtain the best hide for fur, kill the animal in winter, when its fur is thickest for protection and usually is in best condition.

To help the animal survive the cold, its skin often will be depleted to build up the fur. For this reason, the best time to obtain a hide for leather is in summer or early fall, before the skin has been robbed of the nutrients that thicken the fur for the cold weather ahead.

When selecting an animal for its fur, notice the color, texture and thickness of the fur. When selecting an animal for its leather-producing potential, choose one whose fur is not so thick. Assuming the animal is in good health, a thinner coat of fur often means a thicker hide.

TANNING FUR

For several reasons, the beginner may want to experiment first with fur tanning. Suitable small pelts — the hides of squirrels, rabbits and other small animals — are readily available at no cost, and there is no great loss if the first attempts fail. Small amounts of tanning solutions are needed and therefore cost less, while the small size of the hides makes them convenient to handle. Small furs can be very useful, too, for gloves, mittens, hats and other small articles.

A small fur also is a good first project because if at first you don't succeed you can try, try again. If, after it has been tanned and worked, the skin is stiff and hard, it can be soaked or retanned and worked again and again, until it is soft and supple. It is not too difficult to handle a stiff, hard rabbit skin for instance, but a bear skin that is rigid as a board is very hard to manipulate.

When tanning hides for furs, select a thick, good-looking fur to begin with, for it will look no better after it is tanned. The condition of the fur is immaterial if the hide is to be made into leather, since only the skin will be retained. Be careful, too, not to damage or discolor the hair in the skinning or fleshing processes. And don't forget, bark tanning (which gives leather its typical brown, "leathery" color) also will give the fur an undesirable brown, leathery color.

These are the 12 steps necessary to tan skins for fur:

1. Skinning the animal
2. Stretching the skin
3. Cleaning the fur
4. Salting the hide
5. Storing the skin
6. Soaking the hide
7. Fleshing the hide
8. Tanning the hide
9. Rinsing
10. Softening the skin
11. Cleaning the fur
12. Finishing

Those are the steps. When you have followed them all, 1 through 12, you should have a soft, beautiful, tanned fur for your efforts. Ready? Let's start.

Steps to Tanning Furs

STEP 1: SKINNING

The first — and one of the most important steps in successful tanning — is skinning the animal. If the skinning knife is used too close to the skin, the hide may be gouged or torn, making it of inferior quality or possibly worthless no matter how carefully you follow the remaining 11 steps.

However, if the knife is not used closely enough, the hide will have slabs of fat and flesh clinging to it, making the fleshing process doubly difficult.

Only practice will make you expert at skinning, but the judicious use of a good skinning knife and the following suggestions should help:

Small animals such as beaver, raccoon, opposum, muskrat, fox and mink should be skinned by cutting the skin across

A typical skin outline

the legs from heel to heel. Hang the animal by the hind legs and peel the skin off, using a sharp skinning knife to loosen the skin where necessary. Slit the front legs from foot to shoulder and cut off straight at the neck.

The skinning operation should be done as soon as possible after the animal is killed. When hunting, if it is not possible to take the carcass home within a few hours after the kill, it should be skinned in the field and the pelt cooled, salted and folded carefully. Read Step 4 for details.

STEP 2: STRETCHING

To keep the skin from curling and shriveling as it cools, invert it on a drying frame, flesh side out, stretched just enough to keep it from wrinkling without stretching it out of shape.

If you have no drying frame, slit the hide up the middle of the belly and lay the skin out flat, hair side up, pulling it out taut, on a board or piece of plywood large enough to hold its full length and width. Using small brads, nail the left side of the skin to the board, starting at the middle and pointing the front legs forward and the back legs backward, but without stretching the skin in any way.

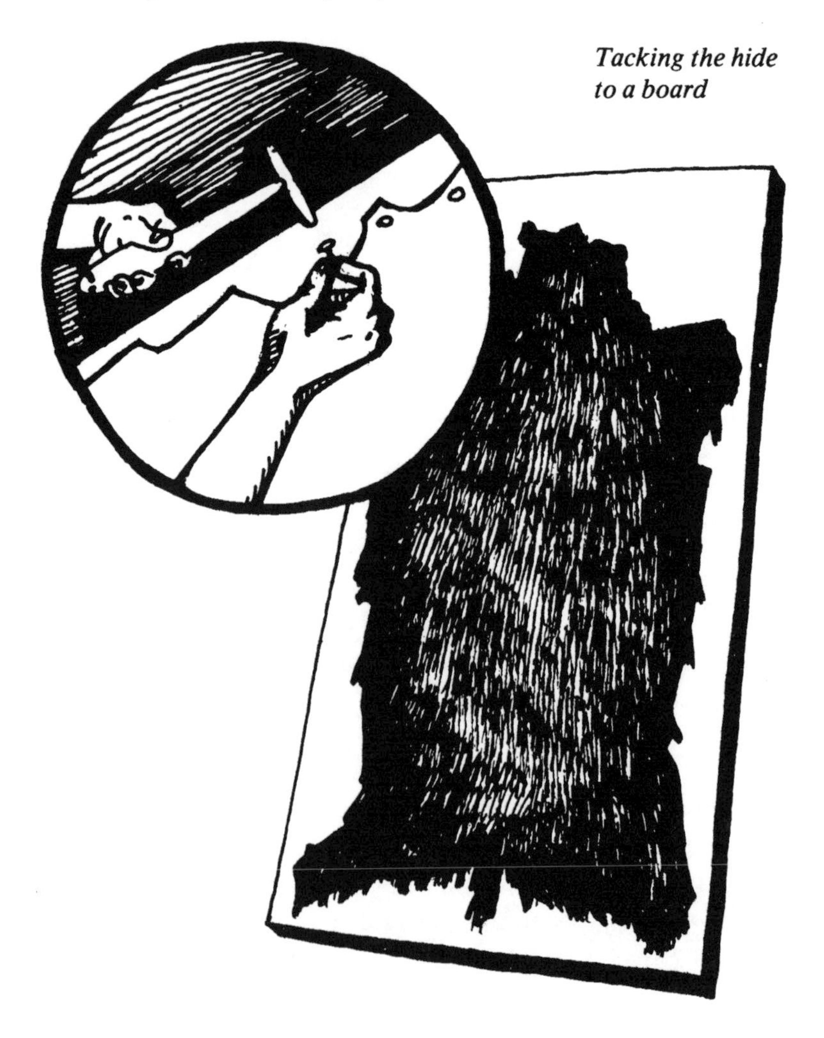

Tacking the hide to a board

Now nail the right side of the skin, keeping the nails exactly in line with the other side and keeping the skin taut but not strained at any point.

This method has the advantage of being inexpensive, even if several skins need to be stretched at one time. It can be done very quickly once you get the knack of knowing when the skin is stretched just tight enough.

Stretching muskrat fur on a wire stretcher. (SCS photo — Herb Kollmorgen)

STEP 3: CLEANING

As soon as the hide is cooled well, scrape off as much of the flesh, fat, dried blood and dirt as possible, using a blunt knife, a kitchen spoon or the dull side of a fleshing knife. It is important at this point to scrape off as much loose material as possible without cutting or gouging the skin. Use only a blunt tool. Small or tightly attached pieces which are difficult to remove may be left. They will be loosened during the soaking process which comes later. Remove from stretcher board or frame and weigh the skin.

Now wash the hide in warm, soapy water, scrubbing it with a stiff brush where necessary to remove blood or dirt. Rinse in clear, warm water. Spread out in the shade until almost dry, pulling and stretching as it dries, or put it back on the stretching frame or board.

STEP 4: SALTING

Spread the partially dried hide out flat, flesh side up. If you did not slit the skin down the belly for stretching, you may need to do so now in order to lay it flat. Pour one pound of salt for every pound of hide, in the middle of the skin and begin rubbing the salt in, working from the center to the outer edges of the skin. Cover every inch of the skin, but be careful not to get salt on the fur side. If you are starting with a small skin which weighs less than one pound, it is sufficiently accurate to figure two cups of salt to the pound, which means you could use one cup of salt for a one-half pound pelt.

When all the salt is rubbed in well, fold the skin in half, flesh sides together, then roll up the fur and place on a sloping surface (such as a drain board) so the salt solution will drain out of the pelt as it forms.

After thirty-six hours, unroll the skin, shake out the old salt, lay the skin flat, flesh side up, and resalt it, again using at least one pound of salt for each pound of hide. Rub it in well, roll up as before and set to drain again.

After forty-eight hours, spread the hide out flat and let dry in a cool, airy place away from heat and out of the sun.

STEP 5: STORAGE

The hide is now salted and is called a "green hide" or a "green salted hide." It may be tanned immediately or it may be held three to five months without damage before tanning it. This is particularly convenient for holding skins over through cold weather or until several skins have been collected to be tanned at one time. Salted hides will not keep well through warm weather and should not be allowed to freeze nor be kept near heat. A holding temperature of 35 to 45 degrees is best. Old-time trappers protected them from insects and vermin by sprinkling the salted hides with arsenic as they stacked them, one hide on top of another, in a cold corner of the cabin.

Skins also may be stored by drying them out completely in a cool, well-ventilated place. Fully-dried skins will keep well through warm weather, but the furs tend to deteriorate. A dried skin that has been stored is brittle and must be thoroughly soaked in clear water until it has softened before it can be handled or folded.

(Note: If you wish to tan a fresh skin immediately and without storing it, you may skip the salting process and instead soak the skin six to eight hours in salt water — one cup of salt to each gallon of warm, soft water. Then proceed with Step 7.)

STEP 6: SOAKING

Prepare a soaking solution of one ounce of borax for each gallon of warm, soft water. Soak the hide in this until the flesh and tissues have loosened. Here an agitator-type washing machine not only is a time- and labor-saver, but it produces a better product. Its gentle agitation action soaks and softens the skin quickly, and speed is important to avoid over-soaking the skin and loosening the hair. If hides are soaked in a pan or barrel, stir and work them with the hands or a paddle occasionally. Four to eight hours should be sufficient.

STEP 7: FLESHING

As soon as the skin is softened and the fat and flesh are loosened, lay the hide across a fleshing beam or on a flat surface, fur side down. Now scrape the flesh side with a fleshing tool or a butcher knife to remove all remaining bits of fat and flesh. For the tanning solution to penetrate the hide, it is essential to remove the tight layer of membrane which completely covers the fleshy side of the hide, and which is like the tough membrane that lines the inside of an eggshell. This membrane must be completely removed to uncover the soft, chamois-like underside of the skin so it can be penetrated by the tanning solution. If it is left on, the membrane will pucker and tighten, forming a hard, stiff outer layer.

During this scraping process, be very careful not to cut or gouge the skin or cut so deeply as to expose the hair roots. Easy does it. Scrape the flesh side of the hide with long, swooping strokes in a regular rhythm. This is a time-consuming task and should not be hurried. The quality of the finished fur

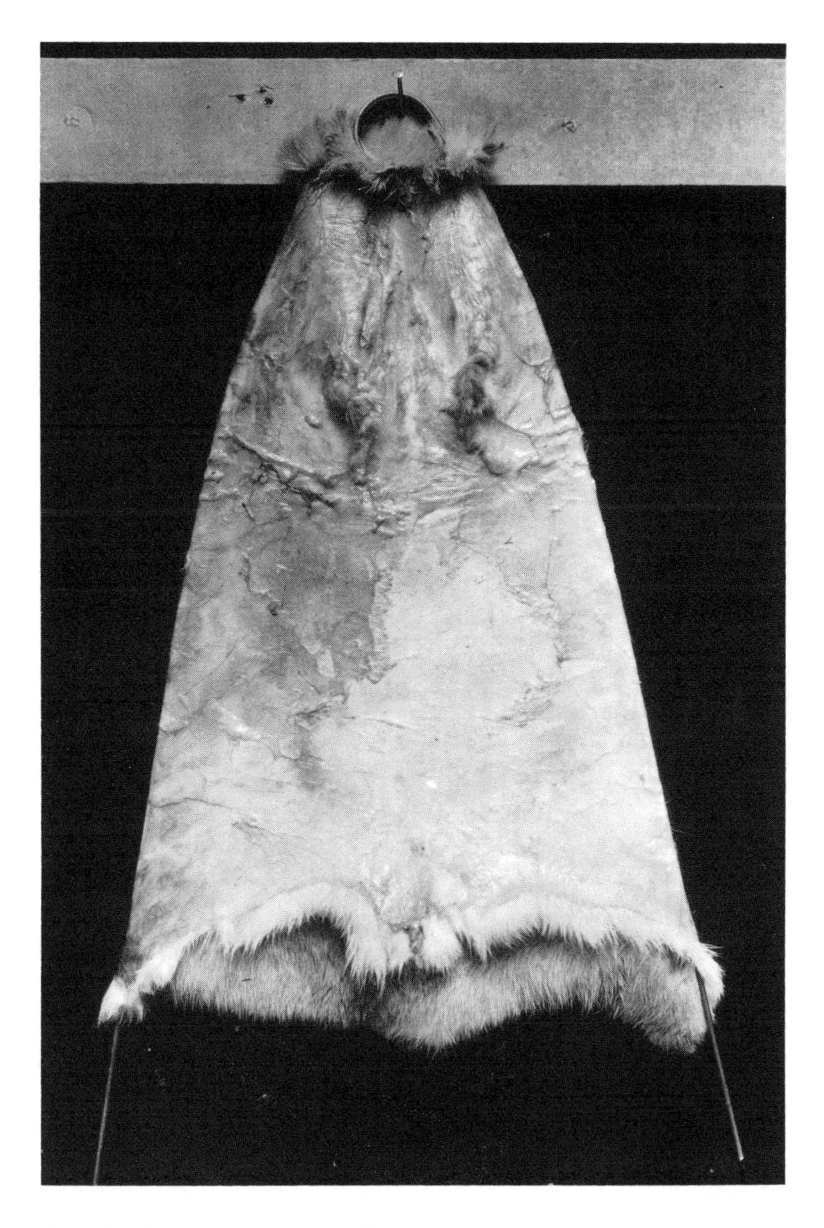

Rabbit skin on stretcher. The membrane shown here must be scraped off during fleshing. (U.S. Bureau of Biological Survey)

will depend to a great extent on the amount of care invested in this step. By using the dull edge of the fleshing tool at least part of the time, you are not only scraping off any excess material, but also are working the hide to help soften it.

Using the beam to flesh a hide

STEP 8: TANNING

This is the step that preserves the fur and skin and keeps it from deteriorating over the years. It also waterproofs it to a certain extent and keeps the skin from becoming stiff and brittle.

There are four basic types of tanning solutions, although the term "tanning" originally referred to the use of a tannic acid solution. Each type has several variations, and you may want to experiment with different solutions to find the type or the variation which best suits your time, your equipment and your choice of finished fur.

Some of the formulas for tanning solutions are for small animal hides. The quantities may be increased for larger skins.

The four basic types are:

A. Vegetable Tanning. This is the oldest of all tanning methods. Old-time tanners soaked hides six months or more in vats of crushed, wet oak bark, until the hides were sufficiently tanned. There is an old recipe for bark tanning on page 53.

Later it was discovered that other barks — especially bark from sumac, hemlock, mimosa and chestnut trees (and galls caused by insect bites on these trees) — also are rich in the tannin which preserves the hides.

Vegetable tannin solutions may be made by grinding tree bark, leaves or the wood of those trees containing tannin. Small furs may be tanned in solutions made of tea leaves, which are rich in tannin. Tannin extracts also are available commercially.

Vegetable tannin is not suitable for tanning fur because it stains the fur. It also is by far the most time-consuming method, but many tanners believe it produces a superior, long-lasting leather.

B. Mineral Tanning. Mineral tanning is preferred by many home tanners and most commercial tanneries because it requires less time and less bulky materials and is far less trouble.

Most mineral processes, however, require far more accuracy

and more care than vegetable processes. It is almost impossible to "overtan" a hide with the bark tanning method, yet it is very easy to ruin a hide by leaving it too long in the chrome or acid solutions.

For this reason, it is important for the novice to read the directions carefully and to follow them exactly, even if there seems to be no reason for such detail. It is important, for instance, to be certain every trace of the chemical is removed by careful washing, in order to stop the tanning process at the right time. Otherwise, the process will continue until the hide is ruined.

C. Oil Tanning. Preserving the hide by oiling is not tanning in the strict sense, but it is included here as an alternative method. It is not used with furs.

Warm oil, usually neat's foot or castor oil, is brushed on or rubbed into the hide, which then is allowed to sit in a warm place while the oil penetrates the skin. Three or four — or more — applications of oil are usually required.

This method is very effective in waterproofing leather, especially leather to be used for boots or harness, where appearance is not important. Although oiling is not so long-lasting as tanning, more oil may be re-applied while the item is being used, so, in effect, it may be "re-tanned" from time to time.

D. Tanning Combinations. There are several procedures that combine two or more of the above tanning methods or one of these with another step added. Indian buckskin tanning is an example of oil tanning plus the added step of smoking the hide.

Some of the recommended methods of tanning furs are listed starting on page 35 under *Formulas.*

Now apply the Test for Tanning (below) and if it shows the tanning process is completed, resume with Step 9.

Test for Tanning

When the hide has been soaking in the tanning solution the minimum length of time called for in the instructions, cut a small slice of skin from the edge and examine it. If the piece shows the same color all the way through, without a lighter layer in the middle, the tanning process probably is complete.

To make sure, drop the small piece in a small pan of boiling water and boil it about five minutes. A piece of hide that is incompletely tanned will curl up and harden. Boiling will not affect a well-tanned piece of hide.

If the piece curls up and becomes rubbery, return the hide to the tanning solution and let it soak, stirring it occasionally, a few days more, then repeat the tanning test.

Lightweight or thinner hides may be tanned in a shorter time and will require smaller amounts of tanning solution than the heavier hides of larger or older animals.

STEP 9: RINSING

Wash out any tanning solution left in the hide by rinsing it in a solution made by dissolving one pound of borax in every gallon of water. Rinse the skin in this solution for ten minutes, stirring and working it with the hands to make sure the borax water removes all tanning solution. After ten minutes, rinse the skin in several changes of clear, soft water.

STEP 10: SOFTENING

Squeeze the water out of the skin and lay it flat, flesh side up. Work over the skin with a slicker to remove most of the water, working from the center, pushing the slicker away from you and working over every inch of the skin. Stretch the skin taut, tacking it down as described on page 21, and apply a thin coating of neat's foot or castor oil. Leave the skin stretched in a cool, dry place, out of the sun and away from heat, until it is almost dry.

While it is still just damp, take up the skin and work the flesh side of the fur over a stake or any rigid wooden surface. Work it vigorously back and forth in a regular, rhythmic action, as though you were putting a high polish on your favorite pair of shoes. This is hard work and requires quite a bit of muscle power, but the suppleness of the finished fur piece will depend on the amount of energy expended at this point. As the fur begins to dry, it may have to be re-dampened repeatedly before the softening process is completed.

STEP 11: CLEANING

By now the fur may be looking pretty soiled and matted from all that soaking and handling, and you may be wondering if it was worth all the work. No matter how soiled it is at this point, however, it can be cleaned and fluffed to its original beauty without too much trouble.

Slight soil may be removed by cleaning with warm sawdust, cornmeal, oatmeal, powdered borax, bran, chalk or plaster of Paris. Warm the dry cleaning material by spreading it out on a baking sheet and putting it in a 250-degree oven five or ten minutes. Then lay the fur flat on a working surface, flesh side down, and work the warmed cleaning material into the fur side. Rub vigorously and work in all the material possible.

To remove the dry cleaner, shake the fur gently over a clean surface or newspaper to collect it (it may be reused unless it is too soiled), then go over the fur with the cleaning attachment of a vacuum cleaner. Brush the fur well, first in one direction, then the other, to remove any remaining cleaning material and to fluff the fur.

When fur has a slight soil with only a few spots of oil or heavy dirt, the spots may be first treated with commercial cleaning fluid or compound, then the whole fur cleaned as above.

Furs with all-over soil may be washed in warm, soapy water; rinsed in clear, soft water and partially dried; then rubbed with the sawdust or other material as described above. Heavily soiled furs may be cleaned in naphtha, benzine or commercial dry cleaning fluids; partially dried; then fully dried as above.

White furs may be washed in warm, soapy water, then covered with a thin paste made of powdered chalk and water. Dry, then brush briskly with a stiff brush to remove all of the chalk.

A homemade, hand-cranked cleaning drum or a cast-off clothes dryer, without the heating element but with a rotating drum, can be a great help with this operation, especially if you are cleaning several furs or a large fur. For each pound of fur (dry weight) measure into the drum a pound of sawdust and rotate the drum about ten minutes, or until the fur is filled with the sawdust. Remove the fur, shake it out well, vacuum and brush it until the sawdust is gone and the fur is clean and fluffy.

A simple, emergency method of cleaning and polishing a fur is to rub it with a piece of fresh bread.

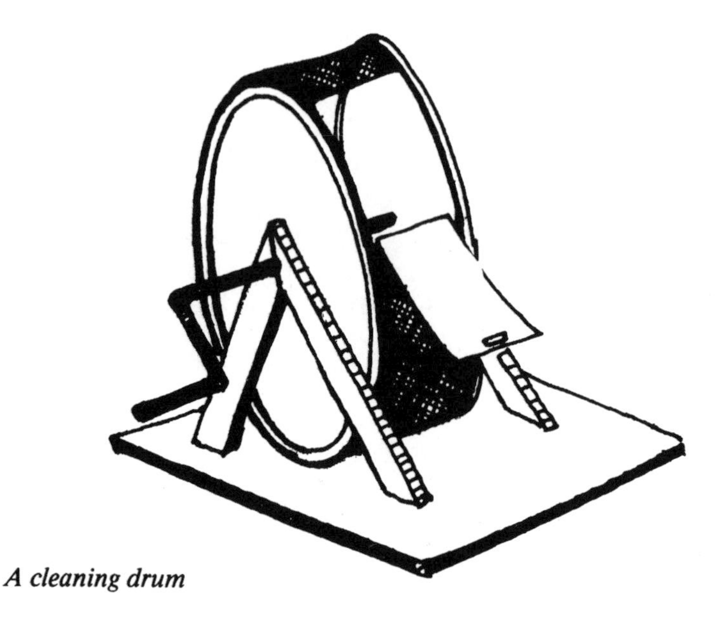

A cleaning drum

STEP 12: FINISHING

Stretch out the tanned fur, flesh side up, on a flat surface. Using fine sandpaper or a damp cloth dipped in pumice stone, remove any rough places, leaving the flesh side uniformly smooth. Shake the fur well, then lay it flat again and apply warm neat's foot oil or castor oil, rubbing it in with the fingers. Apply it evenly over the entire flesh side, using only a very small amount of oil and rubbing it in well with the fingers. Do not allow the oil to get on the fur and do not use too much — just enough to make the leather soft and smooth.

The hide now should be soft and pliable, with a fluffy, natural-looking fur and a soft, smooth underside. Given good care, it will remain attractive through many years of use.

Formulas For Tanning Fur Skins

(Note: Double or triple recipes if needed to cover several or large furs.)

TANNING METHOD #1

Immerse squirrel, skunk, rabbit or other small animal skins in one gallon of warm, soft water in which two cups of salt or two ounces of oxalic acid have been dissolved. (*Caution*! This solution is poisonous. As with all tanning solutions, rubber gloves should be used. Also note the cautions on disposing of tanning solutions on pages 57-58.)

Let soak for 24 hours, stirring occasionally. Larger skins will need a longer soaking time, but do not over-soak so that the hairs begin to loosen.

Continue with Step 9 (page 31).

TANNING METHOD #2

Dissolve one pound of alum in one gallon of soft water. In another container, dissolve four ounces of washing soda and one cup of salt in one-half gallon of soft, lukewarm water. Very slowly pour the soda-salt solution into the alum solution, stirring vigorously as you pour.

Into this solution, immerse the skin which has been cleaned, soaked and fleshed according to the directions in Steps 1 through 7. Soak the skin until it is tanned, no longer than 48 hours for small skins, while larger skins may require three or four days. Stir and squeeze the skin in the solution two or three times a day until a piece of the hide is fully tanned according to the "Test for Tanning" on page 31.

Neutralize the skin by rinsing in a solution of one ounce borax to every gallon of water. Stir well and soak in this solution an hour or more, then thoroughly rinse in clear water. Squeeze out, then stretch the skin, flesh side out, on a stretching frame or according to the instructions on page 21.

Trapper Ivan Anderson with two beaver skins he will tan to make into rugs. (SCS photo — Herb Kollmorgen)

While the skin is still quite damp, apply to the flesh side a thin layer of liquid soap. (This may be homemade lye soap or made by dissolving grated or flaked Ivory or Fels Naphtha soap in a small amount of water over low heat.)

Cool and apply soap with the hand or a paint brush. When this has been absorbed, brush on a coat of neat's foot oil or castor oil, being careful not to get the oil on the fur. Stretch the hide in a cool place, away from heat or sunlight, and allow to dry. When the skin is almost dry, resume with Step 10.

TANNING METHOD #3

To tan small hides into durable furs that will take hard wear — such as furs needed for gloves and mittens — follow directions through Step 7, then spread the flesh side of the skin with a paste made of one part of washing soda, two parts of salt and four parts of alum, mixed with enough soft water to make a thick paste. Apply to the flesh side and leave for two or three days. Scrape off and apply a fresh coating. Repeat three more times, then rinse and proceed with Step 9.

TANNING METHOD #4

Clean and soften the skin according to Steps 1 through 7, stretch the hide out on a flat surface, then spread with a paste made of one pound of salt, one-half ounce of sulphuric acid and a small amount of water. Cover with a sheet of film plastic, leave for six hours, then scrape off the salt and spread the skin with another layer of paste. Leave uncovered and let it dry. Resume with Step 9.

TANNING METHOD #5

Make a salt-acid solution, using one pound of salt for each gallon of soft water. Dissolve the salt in the water, then pour in one-half ounce of sulphuric acid, being careful not to splash the liquid on skin or clothing or to inhale the fumes.

Cool, then add the cleaned softened skins following Step 7, covering them well with the solution. Let soak in the tanning solution one to three days, according to the thickness of the skins. Stir several times a day. When the skins are tanned according to the "Test for Tanning," rinse them well in clear water and squeeze out the surplus. Resume with Step 9.

Now drop the skins in a solution made by adding one pound of borax to each gallon of water needed. Stir ten minutes (a washing machine is convenient for this step), then rinse several times in clear water.

Remove surplus water by working it out with a slicker. Stretch, flesh side out, on a fur stretcher or tacked to a flat surface. Immediately rub with a coating of neat's foot or castor oil, keeping the oil from the fur side. Let dry until just damp, then resume Step 9.

TANNING METHOD #6

Combine 2½ pounds of alum, 1 pound of salt and 1 pound of oatmeal. Mix well and add enough sour milk or buttermilk to make a thin paste. After completing Step 7, lay skin out, flesh side up, and spread with a layer of the paste, being careful to keep it from the fur. Soak the skin in the solution twenty-four hours, or until it meets the "Test for Tanning," page 31. Resume with Step 9.

TANNING METHOD #7

Warm one gallon of rain water and add four cups of bran. Allow to stand in a warm room twenty-four to thirty-six hours, or until it ferments. Then heat to almost boiling and add 2 cups of salt. Cool and add 1 pound of alum. When lukewarm, immerse the skins in the wet bran. Let stand thirty-six to forty-eight hours, then resume with Step 9.

TANNING METHOD #8

Dissolve 3 cups of salt, 2 ounces of saltpeter and 1 ounce of borax in 1 gallon of warm, soft water. Add 1 gallon of sour milk or buttermilk, then stir in eight ounces of sulphuric acid, mixing well, but being careful not to splash the liquid or inhale the fumes.

Put the soaked and fleshed skins in this liquid and stir them every hour or so for three or four hours, then resume with Step 9.

BUTTER TANNING, #9

Wash the hide, if necessary, and allow it to partially dry, scraping as it dries to remove all flesh and fat and the inside membrane.

Spread the flesh side of the pelt with rancid, salted butter, being careful not to get butter on the fur side of the hide.

Over the butter sprinkle a mixture of equal parts of salt, saltpeter and borax which have been mixed together. Fold the skin in half, fur side out, then fold again. Roll it up tightly from the middle corner so that the fur is not soiled by the tanning mixture. Roll it tightly in several sheets of paper and set it in a cool, dry place for two to three weeks.

Unroll, shake off the excess mixture and work the flesh side over a stake until it is soft and supple.

SHORTCUT TANNING, #10

Try this simplified method of tanning for very satisfactory results in preserving furs of small to medium-sized animals:

As soon as the animal has been killed, skin carefully. If fur is soiled, wash in warm, soapy water and rinse well. Lay, flesh side up, on towels in a well-ventilated place out of the sun and allow to dry until just damp.

Rub the flesh side of the skin with a mixture of two parts of salt, one part of saltpeter, one part of alum and one part of bicarbonate of soda (baking soda). Use enough of the mixture to cover the flesh side well, starting in the center and working toward the edges. Cover the edges well, but take care not to get the mixture on the fur side of the skin.

When the flesh side is well covered, fold the skin in half, fur sides out, then fold it in half again. Roll tightly, beginning at the middle of the fur and rolling outward. Roll it up in a newspaper and lay at a slant so that any fluid that builds up will drain out. Place it in a dry, cool place for one to two weeks.

At the end of that time, shake excess moisture and mixture from the skin and scrape with a fleshing tool or a blunt knife to remove any traces of flesh or fat. Being careful not to damage the hide, use the knife to peel off the inside skin-like membrane that clings to the skin.

Vermont farmer with pelts from three foxes he trapped. (USDA photo — Rothstein)

Lay the fur to dry on a fleshing board or table, flesh side up. From time to time pull it lengthwise, then widthwise, to keep it soft and pliable. As it dries, turn it flesh side down over a fleshing stake or a round bar and pull the skin back and forth the way you polish a pair of shoes. When it is dry and soft, the fur is tanned.

TANNING SHEEPSKIN, #11

Soak the hide in cold water 24 hours, or until the skin is soft.

Lay the hide, wool-side down, over a fleshing beam or on a flat surface. Scrape with a fleshing knife to remove all traces of fat and flesh and the top layer of membrane tissue.

Thoroughly wash in warm soapsuds, scrubbing with a brush where necessary to remove spots or foreign matter. Rinse, then let it soak 15 minutes in fresh soapsuds in which ¼ cup of a commercial non-chlorine bleach (such as Clorox 2) has been added. A wringer washer is especially useful for this and for the following step:

Now rinse thoroughly in warm water until the water is clear, and run it through a wringer or squeeze out the water by hand.

While the hide is still damp (directly from the wringer or allowed to dry a few hours if squeezed by hand), spread the flesh side with a mixture of ½ pound of alum, ¼ pound of saltpeter and 1½ pounds of bran. Moisten with warm water and stir to a paste, then completely cover the flesh side of the skin.

Fold the skin, flesh sides together, then roll loosely and let set one week in a cool place.

At the end of that time, scrape off the bran mixture and work the skin until it is soft, by rubbing the flesh side up and down on a metal washboard. Rub every inch of the skin until it is soft and supple.

Tanning Sheepskins with Glutaraldehyde. This method, developed by the U.S. Department of Agriculture, allows a broader use of sheepskins, since when properly tanned this way they are washable in home laundry equipment and are much more resistant to shrinkage than those tanned by the preceding method. This makes the tanned skins most useful for clothing and other varied purposes.

Before tanning, the sheepskin should be washed thoroughly in several changes of lukewarm cleaning water (not above 90 degrees). Use one or two cups of soap or detergent for every 11 gallons of water.

When the skin and wool are clean, drain over a beam or sawhorse, then rinse in several changes of lukewarm clear water. Again drain over beam.

While the skin is still wet, throw it over a beam or on a flat surface, wool side down, and scrape with a fleshing knife to remove any remaining fat, flesh and tissue from the skin. Scrape carefully, using moderate pressure, until only the under layer of the skin remains.

Rinse well in lukewarm water and drain 30 minutes on the beam. Squeeze excess water from the wool and begin the tanning process.

If the skins must be stored first, they may be refrigerated overnight (but be careful they do not freeze) or they may be salted as in Step 4 for longer storage.

Before tanning, weigh the still-damp sheepskin and for each pound of skin, dissolve 10 ounces of technical grade salt in 5 quarts of lukewarm water (85 to 90 degrees) in a clean, watertight wooden barrel. A large wooden, stainless steel or enameled container may be substituted, but do not use galvanized iron or aluminum. Stir with a wooden paddle until dissolved, then add 2¼ ounces of 25 percent glutaraldehyde solution. Stir this in carefully to avoid splashing, and do not inhale the vapors, for the chemical is very irritating.

When the glutaraldehyde is well mixed into the salt solution, carefully immerse the sheepskin and slowly stir it with the wooden paddle for five minutes. Cover with a wooden lid and stir for one minute each hour the first day. At the end of the day the color of the skin and the wool will have turned to a pale yellow. Cover and let it sit overnight.

The second day, again stir for one minute each hour, keeping the barrel covered between stirrings and overnight. Continue this for another six to eight hours — a total of fifty-four to fifty-six hours — then drain the barrel and rinse the skin in several changes of clear, lukewarm water. Lay it over the beam or sawhorse to drain overnight. The next day make a fat liquor emulsion by mixing one cup of neat's foot oil with one cup of water for every twenty-five pounds of sheepskin (based on its original weight). Then mix in two ounces of household ammonia for each pint of oil-water mixture. Stir well and then divide the fat-liquor in two parts.

Place the sheepskin on a flat surface and — using a paint brush or the hand — apply a thin coat of the emulsion to the skin side of the hide, being careful not to allow it to soil the fur. Spread evenly over the entire skin surface. Let it set for thirty minutes and then apply another coat with the second portion. Cover the flesh side with a sheet of plastic to prevent drying, and leave it overnight.

The next day remove the plastic, turn the skin wool side up, and allow it to dry. While it is still damp, stretch it according to the instructions in Step 2. When the skin is almost dry, follow the directions for staking in Step 10. Dampen and stake the skin until it is soft and flexible.

When the staking operation is completed and the sheepskin is soft and dry, the tanning process is completed. To dress the sheepskin, sand the skin side smooth with a sanding block, and comb the wool with a curry comb or wire-toothed dog brush. The wool may be sheared or trimmed to even lengths.

This tanned sheepskin may be laundered in warm (120-degree) water and soap or detergent by hand or in the washing machine. Limit washing time to five minutes to avoid matting, and spin to extract the water. Hang in the shade to dry and repeat the combing process.

TANNING LEATHER

The process of tanning hides into leather is basically the same as tanning furs, except the hair must be removed before the tanning step. A considerably longer soaking time is allowed, since loosened hair roots is not the problem, but is to be encouraged. Remember that thicker hides (which usually are used for leather) require a longer tanning period.

The Fifteen Steps

This is the step-by-step method for tanning leather:

1. Skinning the animal
2. Stretching the hide
3. Scraping the hide
4. Salting the hide
5. Storing the skin
6. Soaking the skin
7. Fleshing the hide
8. Liming the hide
9. De-hairing the hide
10. De-liming the skin
11. Tanning the hide
12. Rinsing the leather
13. Drying the leather
14. Softening the leather
15. Finishing the leather

Skinning on the gambrel

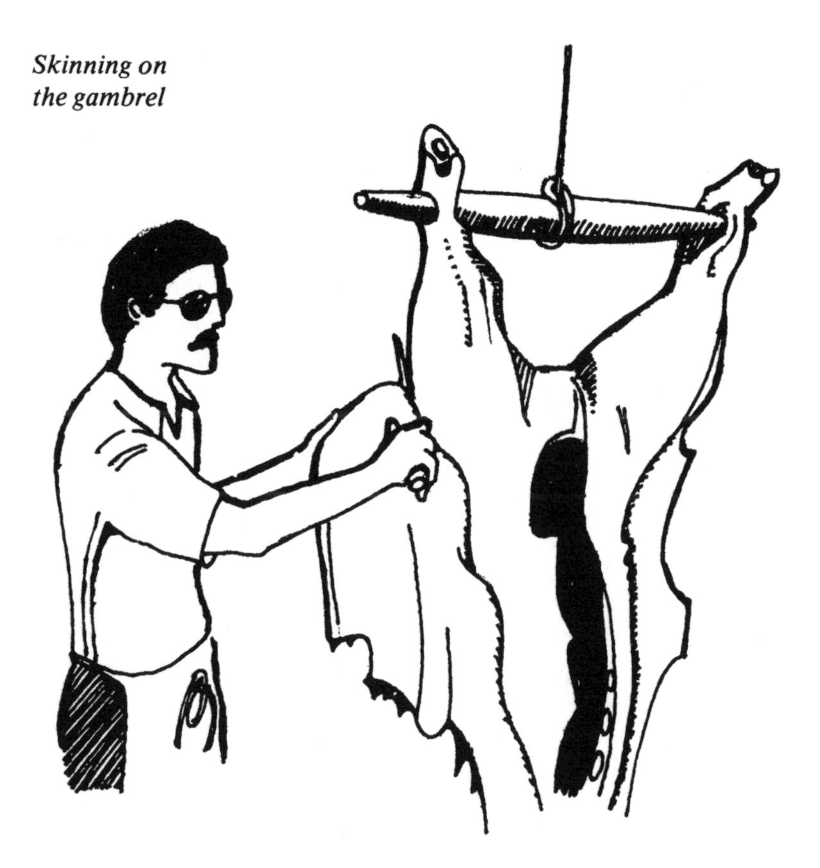

STEP 1: SKINNING

Skin the animal carefully, using a skinning knife to loosen the skin where necessary, according to the directions on page 20.

STEP 2: STRETCHING

Stretch the skin on a stretching frame or nail it taut as directed on page 21, until the skin is cooled of all body heat.

STEP 3: SCRAPING (CLEANING)

Lay the skin out flat on a fleshing beam or a flat surface, flesh side up. Scrape off all excess fat and flesh with a blunt knife or a large kitchen spoon. Unless the hide is very messy, it is not necessary to wash it first, since the hair will be removed and the leather will be soaked and rinsed many times during the tanning process. In the soaking and liming steps, much of the flesh and fiber will be loosened. At this point, the scraping is intended to remove only any excess amounts.

STEP 4: SALTING

When the excess flesh and fat have been removed, weigh the hide to determine the amount of salt to use. Now spread it out flat again, flesh side up, and sprinkle with one pound of salt for every pound of skin. Rub the salt in well, working it right out to the edge of the skin. Since the hair will be removed later, there is no need to be careful of the fur side.

Fold the skin in half, salted side in, and roll it up tightly. Place it on a sloping surface so the liquid will drain away, and let it sit for two days.

At the end of that time, unroll the skin and shake out any loose salt. Then lay it flat again and re-salt the flesh side with another pound of salt for every pound of skin. Rub it in well, roll it up and set it to drain as before.

After two more days, unroll the skin and let it dry, spread flat, away from heat and out of the sun.

STEP 5: STORING

The green, salted hide now may be stored for three to five months without damage. Protect it from insects or small animals, and keep it at a temperature of 35 to 45 degrees. Or, if you prefer, you may proceed with Step 6.

STEP 6: SOAKING

When you are ready to tan the salted hide, scrape off as much salt as possible, then immerse it in a solution of 1 ounce of borax for every gallon of warm, soft water. Let it soak for three to five days, stirring three to four times a day. Hides being tanned for leather may be soaked much longer periods than those being tanned for fur, since loosening the hair by soaking is desirable anyway.

To a point, the longer the soaking time, the easier it will be to remove the flesh, fat and membranous tissue in the fleshing process which comes next.

STEP 7: FLESHING

Remove the hide from the soaking solution and rinse it well. Then throw it over the fleshing beam or a flat surface, fur side down. Using the fleshing knife, scrape every inch of the hide, using long, sweeping motions of the knife. Use enough pressure to remove all the flesh and fat, then scrape away the tough membrane that completely covers the skin.

Work the skin thoroughly, removing every bit of fat and flesh and the inner skin which protects the derma from chemical action. This step will convince you that tanning is more a physical than a chemical process, but the tanning chemicals cannot penetrate the hide evenly unless this under-tissue is removed, leaving only the soft, porous derma layer.

When you are finished, the flesh side of the hide should be completely clean and as soft as chamois.

STEP 8: LIMING

When the skin is soft, pliable and clean of all fat, flesh and tissue, it is ready to be limed.

If you are lucky enough to have a wooden soaking barrel, rinse it out well, almost fill it (about 50 gallons) with soft (rain water is best) warm water, add 8 pounds of fresh hydrated (agricultural) lime and stir well. Immerse the fleshed skins in this solution and keep them completely covered.

Stir and work the hides three or four times a day until the hair comes off easily. This will take from three to six days in warm weather and as long as two weeks in colder weather. *(Note! Always use a wooden paddle to stir skins soaking in such solutions.)*

In a wooden barrel or any other wooden, porcelain, plastic or crockery container (do not use metal). Use the above solution, or any of the three listed opposite in ***Formulas for Loosening Hair Roots.***

STEP 9: DE-HAIRING

When the hair slips easily from the hides — that is, the hair rubs off readily without scraping or pressure — remove a hide

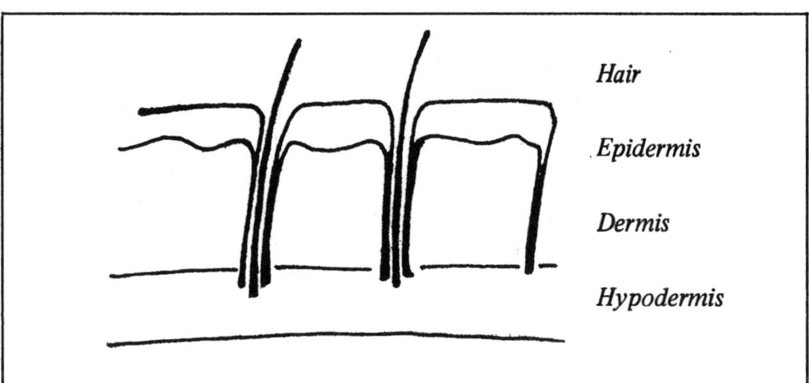

Location of hair roots in skin layers

Formulas for Loosening Hair Roots

Liming Solution #1 (For Leathers)

Dissolve one gallon of hardwood ashes and one gallon of slaked lime in five gallons of warm water. Soak the skins three to five days, or until the hair slips easily.

Liming Solution #2

Dissolve ¼ cup of lye in 10 gallons of warm water. Cool, then soak skins in the solution, keeping them well covered. Stir frequently with a paddle. Soak two days, or until the hair slips off easily.

Liming Solution #3

To every 10 gallons of water, add 2½ pounds slaked caustic lime. Stir with a paddle until completely dissolved. Soak skins in this solution thirty-six to forty-eight hours, or until hair comes off easily.

from the liming solution and lay it across a fleshing beam or on a flat surface, fur side up.

Scrape the hair side of the hide with the dull edge of a fleshing knife to remove all traces of hair and the cheesy layer of the skin which holds the hair roots. The residue from this process — the lime water, hair and any flesh — should be saved. It is a good fertilizer for the vegetable or flower garden or compost pile, or the hair can be used as stuffing for upholstering.

Now turn the hide over and flesh it again, working with smooth, even strokes to clean and soften its flesh side.

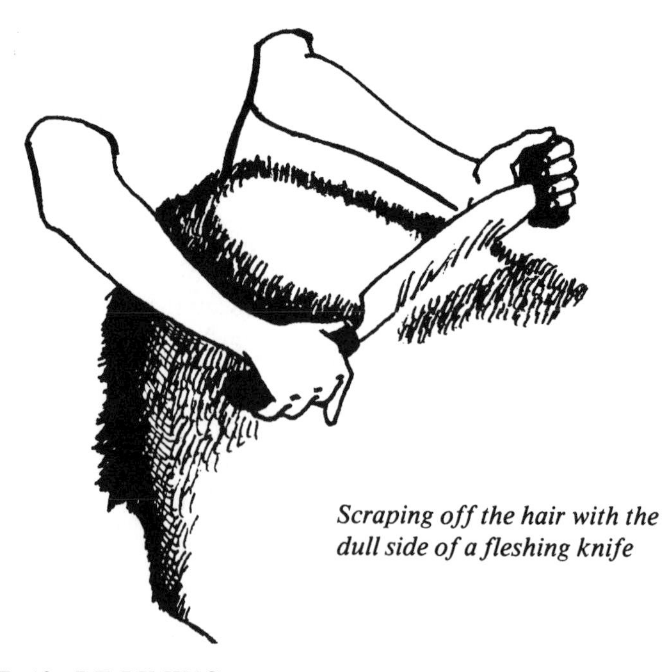

Scraping off the hair with the dull side of a fleshing knife

STEP 10: DE-LIMING

Rinse the de-haired hide in several changes of clean water, then soak it overnight in a barrel or large container of clear water. In the morning, empty the barrel and refill with clear water. Stir in 5 ounces of lactic acid powder or 2½ quarts of vinegar. Mix well. Soak the hide in this solution twenty-four hours, stirring every few hours.

At the end of this soaking time, empty the barrel, re-fill it with clear water and soak the skin overnight.

STEP 11: TANNING

Here are some of the methods for tanning leather:

Bark Tanning. Bark tanning is one of the earliest methods of preserving animal hides, but it is seldom used today because of the length of time and the large amount of bark required. It is also a very messy operation.

However, bark tans leather well and gives it that distinctive "leathery" color. It leaves sole leather very water-resistant and durable. In addition, it is almost cost-free.

Bark tanning infusions must be prepared two to three weeks before the tanning process begins. To prepare an infusion for a large hide or several small hides, finely grind 30 to 40 pounds of oak or hemlock bark by putting it through a grist or hammer mill with the finest screen. If you have an old hammer mill that can be attached to a tractor's power take-off, it would be a great work-saver for this process. A hand-cranked steel grist mill will do the job, but will require a great deal of manual labor.

Pour the ground bark into a wooden barrel or plastic container. Do not use one of iron. For this amount of bark, a sixty-gallon barrel works well. Cover the bark with twenty gallons of hot water and let it stand, stirring occasionally, for two to three weeks.

When you are ready to tan the hide, pour the bark infusion

into a tanning barrel, straining it through coarse material to remove the ground bark. Fill the barrel with soft (rain) water and add two quarts of vinegar. Stir well.

Immediately after the de-liming process (Step 10), immerse the hide in this tanning liquid and let it stand in the solution ten to fifteen days, stirring three or four times a day. At least once a day, remove the hide from the solution and change its position in order to get an even color.

Immediately after you first put the hide in the liquid, set another thirty to forty pounds of ground bark to soak in another twenty gallons of hot water in the wooden barrel. Let it steep, stirring occasionally, for ten to fifteen days, or until the hide in the first solution is evenly colored.

When the hide looks evenly colored on the surface, take the skin from the tanning barrel. Remove five gallons of the old tanning solution from that barrel and discard it. Replace the five gallons with five gallons of the new tanning liquid being prepared in the wooden barrel, straining it as it is added. Add two quarts vinegar and stir well. Replace the hide in the tanning barrel and move it around well. Let it remain for five days, stirring every day.

At the end of this time, remove five gallons more of the solution in the tanning barrel, discard and replace with another five gallons of the new solution. Do not add vinegar this time. Place the hide in the barrel. Repeat this step every five days until the liquid in the wooden barrel is used up.

Continue to soak the hide two more weeks, then remove it and pour another forty pounds of bark directly into the liquid in the tanning barrel. Stir well, then put the hide back, still stirring. Keep the hide completely covered by the bark and liquid another four to six months, or until the "Test for Tanning," (page 31) shows the tanning process is complete. Continue with Step 12.

Chrome Tanning. At least forty-eight hours before it is to be used, make a tanning solution as follows:

For a medium-sized hide, dissolve 1¾ pounds sodium carbonate crystals and 3 pounds of salt in 1½ gallons of warm, soft water. Use a wooden barrel or plastic container. Stir well to dissolve.

Meanwhile, in another wooden container, dissolve 6 pounds of chromium potassium sulfate crystals in 4½ gallons of cold water. This will dissolve slowly and must be stirred constantly.

When there are no crystals in the bottom of either container, combine the two liquids by slowly pouring a thin stream of the first solution into the second. This should be done so slowly that it will require about ten minutes to complete, and it should be stirred constantly during this time.

When the hide is ready to tan, pour one-third of the combined solution into a wooden or plastic container and add 15 gallons of clean, cold water. Stir well, then add the hide, working with a paddle to give an even color. Repeat this agitation every hour or so the first day, then several times a day for the next two days.

At the end of the third day, remove the hide from the solution and add half of the remaining soda-chrome solution, thoroughly stirring it into the liquid in the container. Replace the hide, working it around with the paddle three or four times a day.

After three more days, repeat this process, using all the remaining soda-chrome solution.

Three days later, cut off a small sliver of the hide and test for tanning according to the instructions on page 31. If the tanning is completed, resume with Step 12. If not, return the hide to the container for a few more days, then test again.

Salt-Alum Tanning. (Note: Although this method of tanning is much faster and easier than some of the others, leather tanned with alum is more likely to dry stiff and hard. Although the results can be quite satisfactory, leather tanned this way requires a greater amount of work during Step 14 — Staking.)

Dissolve one pound of alum in one gallon of warm water. In another container, dissolve 2½ pounds of salt in four gallons of water. Slowly pour the salt solution into the alum solution, stirring constantly. Allow it to cool, then immerse the hide. Soak from two days for small, thin skins and up to six days for large, thick hides. Test according to the instructions on page 31.

When the skin is completely tanned, proceed with Step 12.

Alum-Carbolic Acid Tanning. For a large hide, make a soaking solution of 1½ tablespoons of carbolic acid crystals for every gallon of water. Soak the hide in this overnight or until hide is soft and pliable.

Meanwhile in another container, dissolve ½ pound of salt, ¼ pound of alum and ½ ounce of carbolic acid crystals in every gallon of water needed. Use warm water and stir well until all is completely dissolved, then let it set until cool.

Immerse the hide in this solution and let soak until the "Test for Tanning" shows the process is completed. This may take up to six days or more for large hides.

Resume with Step 12.

Sulphuric Acid Tanning. (Caution! Handle sulphuric acid with care. Do not let it touch skin or clothing. Do not splash it in the eyes. Do not inhale fumes. Wear rubber gloves and safety glasses and use a glass, wooden or earthenware container. *Never use metal.*)

Dissolve ½ pound of salt in every gallon of water. Very slowly, very carefully, pour in ½ ounce of sulphuric acid per gallon of water, stirring constantly and steadily. Cool, then add the hide, moving it around in the solution to wet all parts of it. Soak one to three days, depending on the thickness of the skin. Stir and move the skin around several times a day.

Continue with Step 12.

Acid-Oil Tanning. Make a solution by dissolving ¼ pound of salt in ½ gallon of warm water. Cool, then gradually — very carefully — pour in ¼ ounce of sulphuric acid, stirring carefully so as not to splash the liquid.

Spread out the hide, flesh side up, and paint with a coating of this solution and sprinkle it with a layer of sawdust. Cover with a thin sheet of plastic to keep it from drying out and leave it overnight, or 8 to 12 hours.

Scrape off the sawdust and apply a coating of neat's foot oil which has been mixed with an equal part of water and warmed to lukewarm.

Stretch the skin until dry, then dampen with a liquid made by dissolving 1½ tablespoons of carbolic acid crystals in 1 gallon water. Roll up and let set 24 hours, then resume with Step 12.

STEP 12: RINSING

When the tanning process is completed, pour out the tanning solution in an area away from vegetation and water supplies. *Use caution!* Most tanning solutions — unlike the liming solution — will kill vegetation and may poison animals or people if poured into or near a stream or well. It also is bad for septic tanks.

The decision on whether or not to re-use the tanning solution is largely one of judgment. Sometimes, when only one hide or a few small hides have been tanned, a second use of the solution may be entirely satisfactory if it is used immediately. Or perhaps a longer tanning period the second time would enable you to use the solution again.

With some solutions — such as that used for chrome tanning leather on page 61 — it is not possible to use the solution again because the chemicals are added gradually to strengthen the liquid. With all solutions, because of the danger as well as the deterioration of the chemicals, it is best not to store them for the next tanning operation. In general, it is better to store the hides until you have several to tan at once, and then begin each tanning operation with a fresh solution.

When the tanning solution is discarded, clean the tanning container and refill it with water, in which 2 pounds of borax has been added for every 40 gallons of water.

Meanwhile, thoroughly rinse the hide in running water, then immerse it in the borax solution and let it soak overnight. In the morning, once again rinse the hide in several changes of clean water.

STEP 13: DRYING

Now pull the hide taut and nail it to a flat surface, as described on page 21, or put it on a stretching frame to dry.

While it is still wet, wipe the hair side with a coating of warm neat's foot oil. When the hide is just damp, wipe off any excess oil and remove the hide from the stretched position.

STEP 14: SOFTENING

Staking is a method of working the tanned hide back and forth over the edge of a hard, smooth surface until it is supple and softened. Unless staking is done, or if it is improperly or incompletely done, the leather will be stiff and brittle. It is hard work, but it is the only way to make leather soft and supple.

The process is simple. Just rub the leather piece back and forth, back and forth, across the stake, using a certain amount of pressure and with a fairly brisk rhythm, much as you would to polish a pair of shoes on a boot black stand.

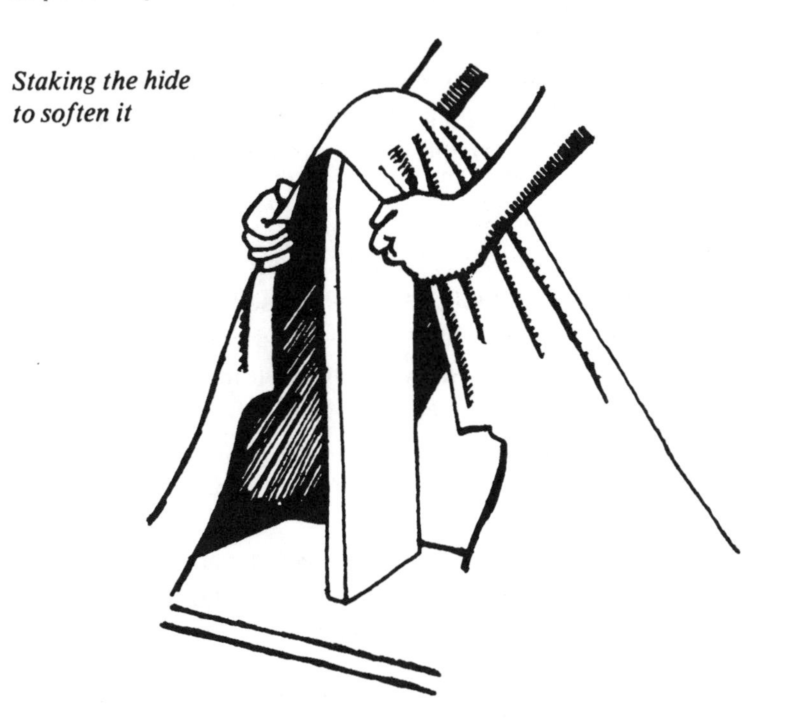

Staking the hide to soften it

While it is damp, every inch of the leather should be staked in this way until the entire piece is soft. If the hide dries, it should be dampened by wiping it with a well-moistened cloth or by placing it for a few hours between damp towels.

STEP 15: FINISHING

When the leather has been staked until it is soft and pliable, dampen it again, and then apply a coating of warm dubbin to the hair side.

To make dubbin, melt 1 pound of beef or mutton tallow and combine with 1 pound neat's foot or castor oil. Cool, then apply it to the hide with a paint brush or the fingers. Let it set in a warm place until dry, then stake again to soften the leather and work in the dubbin.

Lastly, rub the hide on both sides with fine sandpaper or a damp cloth dipped in pumice stone to smooth the surface. Give it a final coat of dubbin and wipe it off with a soft cloth.

That's it. You now have a piece of leather. If you have followed all 15 steps carefully and have not stinted on the fleshing and staking steps, it should be a durable piece of leather.

It may not look professionally tanned. It may not be the best piece of leather you've ever seen. But it is a piece of leather you tanned yourself. In the process, you not only transformed a piece of useless hide to useful leather, you also learned a great deal about a very old craft.

The success of any tanning project depends upon three variables — the temperature, the chemicals and the experience of the operator. The next time you tan a hide, the results will be better, because even if the first two variables are precisely the same, the operator will be more experienced.

To Tan Sole Leather

Leather for belts and shoe soles is made of heavy hides from large, mature cattle and horses.

Because of its thickness and the resulting weight, and because sole leather should be more rigid than leather for other uses, these hides are handled in a different way.

The heavy hide is first divided down the middle, lengthwise, to make it easier to handle. Each piece is then weighed and salted as in Step 1, using one pound of salt for every pound of hide. The hide pieces are then stacked, flesh side up, and stored forty-eight hours or more, (or they may be stored in a cool place for several months).

The two sides of the hide are prepared for tanning by soaking in clean water for eighteen to twenty-four hours, or until the hide is soft and pliable.

They are then fleshed thoroughly, using long, smooth strokes and sufficient pressure to remove every trace of fat, flesh and membranous tissue.

In a sixty-gallon barrel, add ten pounds of hydrated lime and stir until well dissolved. Immerse the two pieces of hide in this and let soak ten to fifteen days, moving them about several times a day.

When the hair slips off easily with little pressure, take the hides from the limewater and lay them over a flat surface or fleshing beam. Scrape off the hair and the top, cheesy layer of skin, then turn each piece over and again go over the flesh side.

To remove all traces of lime, rinse the hide in several changes of water, then soak twenty-four hours in a barrel of clear water in which 5 ounces of lactic acid or 2½ quarts of vinegar have been dissolved.

Meanwhile, make a tanning solution by dissolving 6 pounds of salt and 3½ pounds of crystallized sodium carbonate in 3 gallons of warm water. In another container add twelve pounds of chromium potassium sulphate crystals to 9 gallons of water and stir until dissolved. Add to this the salt-soda solution very

Skinning the heavy hide from a cow prior to tanning. (USDA photo — M. C. Audsley)

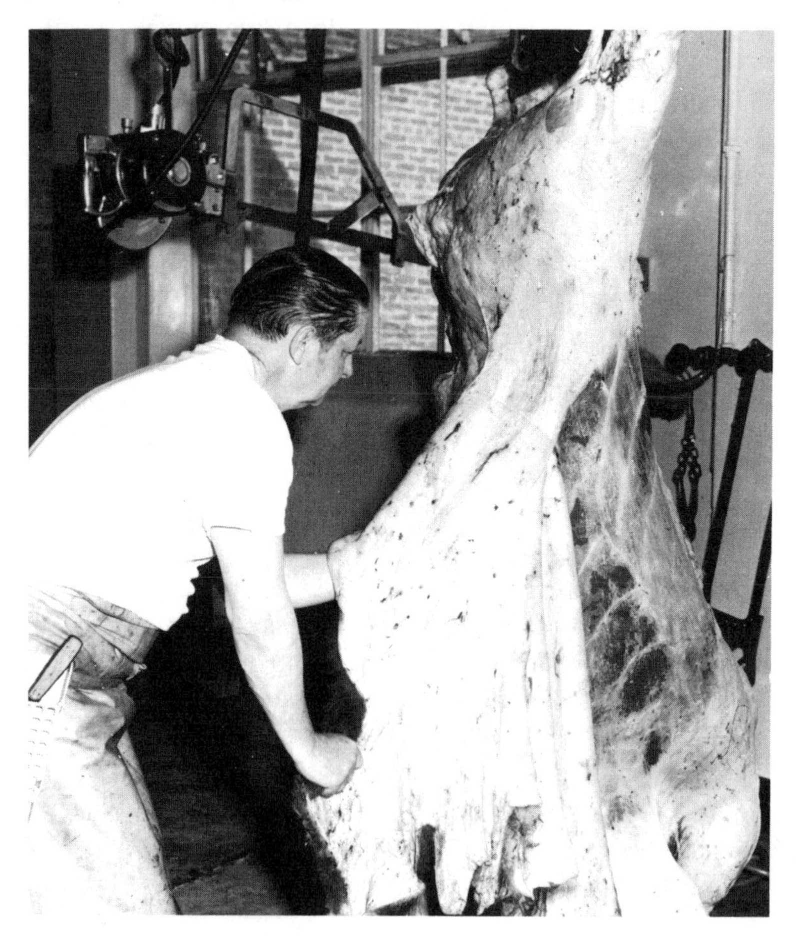

slowly, stirring constantly. Measure out four gallons of this combined solution and mix it with forty gallons of clear water in the tanning barrel. Immerse the hides in this and let them soak for three days, moving them about several times a day.

After three days, remove the hides, add another 4 gallons of the original combined solution and stir well, then replace the hides. Let them soak another three days, stirring often. At the end of that time, add the last of the combined liquid in the first container and replace the hides. Let them soak another three days.

At the end of this time, test the hides for tanning, according to the instructions on page 31. If the process is not completed, return the hides to the barrel another three days, then test again. If tanning is complete, rinse the hides well in several changes of clean water, then soak eight to twelve hours in a solution made by dissolving two pounds of borax in every forty gallons of water needed. Remove from the solution and rinse them well in clean water.

Heavy leather is then coated on both sides with a liberal dressing of neat's foot oil and let set one-half hour. It is then slicked with a slicking tool, a wedge-shaped piece of hardwood about six inches wide. Stretch the hides out flat, then slick each piece from the center to the outer edges, using a great deal of pressure to remove much of the water and excess oil and to smooth out the surface. Continue until the leather is smooth and slick. You may have to sprinkle it with additional warm water as it dries.

Because of its heavier weight and greater thickness, and because softness is not desired, sole leather is not staked, as are other types of leather.

Indian Buckskin

Long ago, under very primitive conditions and without a controlled environment or special equipment, the American Plains Indians made a soft, supple buckskin using only the tools and supplies they had on hand while hunting. Here is the method they used:

After carefully skinning a deer, they filled the scooped-out hollow of a large log with clear rain water, into which they threw a handful of wood ashes to make a dilute lye solution. They soaked the deer hide in this until the hair slipped off easily.

The skin was then thrown across a peeled log and scraped with a thigh bone of the animal. By long, rhythmic scraping, the hair was removed from one side of the skin and all fat and membranous tissue were removed from the other.

Meanwhile, the animal's brain and a handful of tallow from the body were mixed with hot water and cooked over the open fire, then allowed to cool. When the scraping with the thigh bone was completed, the brain-tallow mixture was rubbed well into the skin with the hands. Next the skin was allowed to dry partially.

The hide was then pulled over a sharpened stump and scraped with stones and shells until it was soft and supple and of uniform thickness.

The skin was then smoked over a slow-burning fire of half-rotted wood, a process that gave the cured skin a tan color, a certain water-repellent character and a pleasant odor. A tent-like structure of branches was made over a trench which led to the covered fire. The skin was smoked three to five days, or until it was a golden tan color.

MODERN BUCKSKIN TANNING

Although buckskin is not as waterproof or as long-wearing as leather which has been preserved in tanning solution, it can be beautiful and useful.

The advantage is that it can be made quickly and easily, with few special tools and in only a few days' time, while the more durable tanned leathers take several weeks and some equipment. Here is a method of producing a tough, durable buckskin similar to that made by the Indians but under more modern conditions:

Cool the freshly-skinned hide of a deer, elk, antelope, wolf, sheep, calf or goat overnight, or until the body heat is gone. Put it to soak in a solution of ¼ cup of lye dissolved in ten gallons of cold water. Soak the skin overnight or until the hair will slip off easily without pulling.

Spread the hide on a fleshing beam (or peeled log) and scrape with a fleshing tool or butcher knife to remove all the hair. Rinse off, turn the hide over and work with the fleshing tool to remove all traces of fat, flesh or membranous tissue.

Soak the hide again overnight in a sufficient quantity of diluted vinegar to cover, using two cups of vinegar to every ten gallons of water. In the morning, rinse thoroughly and immerse in the following solution:

Dissolve two cups of homemade soft soap (or two bars naphtha soap which have been dissolved in one cup hot water) in three gallons of hot water. Cool, then put the hide in this and allow it to soak, completely submerged, four or five days, stirring it several times a day.

Remove the hide from the water, rinse in clear water and pull and stretch it until it is partially dry. Smear the hide generously with a coating of lard, bacon grease or neat's foot oil and immerse it in a new solution of warm soap suds made as before. Let it soak three days more, stirring daily. Rinse it again and pull and stretch and work it over a stake as it dries. When the skin is partially dry, tack it to a frame, stretching it taut. Allow it to dry in a cool, airy place away from sun and heat. When it is partially dry, work the skin over a stake until it is soft and supple.

To smoke, drape the skin tent-like over a framework of poles set over a very small fire, smothered with damp wood to keep it smoking. Or use a smoke house or one improvised by placing a barrel over a trench leading to a fire or fireplace. Smoke until the buckskin is a golden tan.

In making buckskin, it is important to keep working the skin as it dries. If the skin begins to dry stiff and hard, it should be dampened with wet cloths and reworked.

Buckskin will stiffen after being wet. It can be softened again by working it well while it is still damp.

Tanning Snakeskin

Skin the snake first by cutting off the head and slitting the skin from end to end down the middle of the underside. Peel off the skin, handling carefully to keep from tearing, and using a skinning knife where necessary.

The skin can be tanned immediately or it may be soaked in a salt-borax solution (½ pound of borax and 3 pounds of salt in 1 gallon warm water). Let it soak for forty-eight hours, stirring occasionally, and then hang it to dry. When dried, it may be stored for several months, but do not allow the storage room temperature to go below 40 degrees or above 60 degrees.

To tan the salted snakeskin, soak it for three to five days in a weak lime solution (3 ounces of slaked lime to 1 gallon of warm water). Remove the skin from the limewater, spread out on a flat surface and remove the scales by scraping against the grain as you would scale a fish.

Soak the scaled skin twenty-four hours in a boric acid solution (1 ounce of powdered boric acid dissolved in 1 gallon of warm water). Rinse the skin in several changes of clean water.

In a wooden or plastic container, make a tanning solution by dissolving 15 grams of chrome alum and 4 ounces of salt in 1 gallon of warm water. Stir well until thoroughly dissolved, and then immerse the skin. Soak it three to five days, according to the size and thickness of the skin, stirring several times a day. At the end of that time, remove the skin from the solution, then dissolve 5 grams of sodium carbonate (*not* bicarbonate) in 1 cup warm water and very gradually add it to the tanning solution, stirring well as you pour in a thin stream.

When completely mixed, put the skin back in the combined solution and stir it well. Let it soak five to seven days, moving it about several times a day.

At the end of the tanning period, remove the skin, drain it and soak it in 1 cup neat's foot oil which has been mixed with 3 cups warm water. Leave it for eight to twelve hours, then remove and drain it. Wipe off the skin and pull it out to dry, tacking it down to pull the skin taut, but not stretched. When it is thoroughly dry, work it gently until supple and softened, being careful not to tear the delicate skin.

When the tanning is completed, cover the snakeskin with a pressing cloth, then press with a warm iron to flatten it. Spray the hide with a clear plastic spray for a gloss.

TREATMENT FOR LEATHERS & FURS

1. Dyeing Leather

Leather may be dyed with oil stains, water stains, drawing ink, or with commercial leather dyes made specifically for this purpose.

Before being dyed, the leather must be cleaned thoroughly to remove any traces of the tanning chemicals. Wipe the leather well with a cloth dampened with a solution of one part hydrochloric acid in twenty parts water.

Before dyeing the pieces of leather, first test a small scrap with the stain or ink or dye to see if the effect is what you want. Two coats may be applied for a darker tone.

If you are satisfied with the test piece, you are ready to dye the article. Brush on the stain or ink as smoothly as possible, without overlapping, and applying carefully to the cut edges. Dry thoroughly, then buff with a soft cloth.

Leather may be painted with commercial shoe dyes available in drug and variety stores, or it may be tinted by immersing it in a concentration of commercial fabric dye, then rinsing it in clear water.

MAKING YOUR OWN BLACK DYE

Leather may be dyed black with iron liquor and an extract of sumac leaves. To prepare the stain, let 2 cups of unrusted iron filings stand in ½ gallon of vinegar for ten days to two weeks, adding filings as needed.

Meanwhile, put 10 to 15 pounds of dried sumac leaves in the tanning barrel. Add 40 gallons warm water and stir well. Let cool, then add the leather to be dyed.

Let it soak for two days, then remove the leather from the solution of sumac leaves. Rinse and paint it with the water in which the iron filings have been steeped. Rinse again, then return to the sumac barrel and let it soak another twenty-four hours.

If the leather is black enough at this time, rinse well in several changes of water, then hang to dry, working to soften it as it dries. If the leather is not black enough, again paint with the iron liquor, rinse, and then immerse in the sumac liquid for two more days.

2. Dyeing Furs

Fur may be dyed simply and safely with commercial hair dyes, following manufacturer's directions. Apply with a toothbrush or for special effects, use a small pointed-tip paint brush. It is always a good idea to practice first on a small, worthless skin.

3. Preserving Leather

Leather that is used often or is exposed to water — such as leather boots or saddles or shoes — can be preserved by rubbing with equal parts neat's foot and castor oil.

Using a swab or a soft cloth, spread a layer of oil uniformly over the leather, then rub until it is completely absorbed. Let set 24 hours, then repeat.

This oiling should first be done before the article is used, then repeated at least once a year, according to the amount of use.

Waterproofing Shoes

To protect shoes with a waterproofing wax, smooth the leather by rubbing it with fine sandpaper or a damp cloth dipped in pumice stone, then apply an even coat of wax made by melting together equal parts of beeswax and mutton tallow. Cool and allow it to harden, then mix to a soft paste with neat's foot or castor oil. Mix in lampblack, ivory black or stove blacking to an even black color. This formula also may be used without the blacking to produce a waterproof finish in a natural, brown color.

Apply smoothly, rubbing in well. Allow shoes to dry, then polish softly. Apply several coats to build up a waterproof finish.

Other mixtures which may be used as a leather waterproofing finish in the same manner are:

Equal parts of melted mutton tallow and beeswax.

Four ounces of beef tallow, 2 ounces of beeswax, 1 ounce of rosin, 2 ounces of neat's foot oil and 1 ounce of lampblack.

Eight ounces of raw linseed oil, 4 ounces of beef tallow, 3 ounces of beeswax, ½ ounce of rosin, and 2½ ounces of turpentine.

Two ounces of mutton tallow, 6 ounces of beeswax, 2 ounces of soft soap, 2½ ounces of lampblack, ½ ounce of powdered indigo. Dissolve over low heat, stir well and add ½ cup of turpentine.

Four ounces of raw linseed oil, 5 ounces of boiled linseed oil, 4 ounces of beef tallow and 4 ounces of beeswax. Melt and mix well.

To Deodorize Furs

Bad odors in hides, as may remain in skunk fur, can be removed from animal hides by immersing them in a solution made by dissolving two pounds finely grated lye soap and two pounds of salt soda in every gallon of hot water needed.

Stir to dissolve completely, then add ½ ounce borax and cool to lukewarm. Add ¼ ounce of oil of sassafras, mix well and soak the skin in this liquid one-half hour, working with the hands occasionally. Rinse and dry, then comb and fluff the fur.

TWO VERY OLD TANNING RECIPES

1. For Fur

To tan an undressed hide into fur, spread the clean, cooled skin, flesh side up, on a flat surface and apply equal parts of saltpeter and powdered ammonia. Sprinkle evenly over the surface. Fold the skin in half, fur side out, and roll up. Let stand three or four days until powder is dissolved. Lay flat and scrape the flesh side well, removing all vestiges of flesh, fat and membranous tissue. Nail the skin, flesh side out, on the shady side of a building or a board fence, stretching tightly as you nail. Liberally apply warm neat's foot oil with a brush. Let stand twenty-four hours in a warm place, then work with a wooden wedge. Work with a smooth, rhythmic motion, using as much pressure as possible to remove the excess oil.

2. For Leather

Immerse the skin in a thin milk of lime (a weak solution of agricultural lime and water), for four or five days, then scrape off the loosened hair and any flesh remaining. Wash to free it from lime, changing the water several times.

Plunge the skin into a solution of bran and water and let it stand two days. In another container, boil ½ pound of apple galls, 1½ ounces of Bengal catechu and 5 pounds of torrentil or seplfoil. Soak in 17 gallons water for one hour. Take the skin from the bran water and plunge into this liquid.

Work with a paddle to soften it several times a day for three or four days. Let it stand undisturbed three or four days, then work it three or four times the next day. Let it lie, undisturbed, for one month to complete the tanning process.

SOURCES FOR CHEMICALS, MATERIALS, & TOOLS

Cumberland General Store, LLC
Alpharetta, Georgia
800-334-4640
www.cumberlandgeneral.com
Books on tanning and country living

Cumberland's Northwest Trappers Supply, Inc.
Owatonna, Minnesota
507-451-7607
www.nwtrappers.com
Books on tanning and tanning kits

Fiebing Company
Milwaukee, Wisconsin
800-558-1033
www.fiebing.com
Tanning supplies

Leather Unlimited Corp.
Belgium, Wisconsin
800-993-2889
www.leatherunltd.com
Tanning kits and leather goods

Lehman's
Kidron, Ohio
888-438-5346
www.lehmans.com
Books on tanning and country living, butchering equipment, cutlery, and high-quality old-fashioned tools

PFAU Industrial Animal Oils
George Pfau's Sons Company, Inc.
Jeffersonville, Indiana
800-732-8645
www.pfauoil.com
Liquoring ingredients and Peacock neatsfoot oil

Quil Ceda Leather Company
Marysville, Washington
866-852-9581
www.quilcedaleather.com
Custom tanning and leather goods

Sterling Fur Company
Sterling, Ohio
330-939-3763
www.sterlingfur.com
Complete tanning and skinning supplies

The Tannery, Inc.
Lander, Wyoming
800-866-4094
www.thetanneryinc.com
Tanning kits and leather

Woodcraft Supply LLC
Parkersburg, West Virginia
800-225-1153
www.woodcraft.com
Draw knife

PART II

WORKING WITH LEATHER

by Steven Edwards

LEATHER GOODS FROM HOME-TANNED LEATHER

Introduction

In reading Phyllis Hobson's directions for tanning in the first part of this book you have learned, if there was ever any doubt, that the tanning of leather from animal skins requires a tremendous amount of work. And yet, there are many sound reasons for tanning at home, including the low dollar cost for finished leather or fur, and the satisfaction of having taken hunting or trapping one step further than is commonly done.

And perhaps the most important reason is that home tanning brings one closer toward using all the products from an animal which was hunted and killed — helping to divorce that act from any pure "sport" association, and giving it more ecological and moral justification, in that little is wasted, and the goal of hunting becomes self-sufficiency.

The logical next step for a hunter and home-tanner is to use leather he has tanned to fashion useful articles. If you have invested the time and trouble to tan leather, you are well on your way to total involvement with something you make yourself, from scratch. It is the same sort of involvement which is felt by a woodworker who cuts a tree, mills the lumber and builds a fine piece of furniture. That furniture and your leather articles become personalized extensions of the one who crafted them.

There are many useful articles of clothing and accessories which may be made of leather. The purpose of this part of *Tan Your Hide* is to give information on how to make various articles from the leathers you have tanned using Phyllis Hobson's directions, although the same skills are applicable to commercial leathers.

As Mrs. Hobson has explained, leather can be tanned in several ways, and leathers, even from animals of the same species, can be quite different, depending on many variables: the age of the animal, the time of year of the skinning, as well as the type of tanning agent and finish used, and how they are employed. Skins can be fur skins, finished, sueded, dyed, left undyed, waxed, oiled, and so on. It certainly sounds confusing.

But take heart. A use can be found for any leather you tan, for leather goods are made from all weights and types of leather. The skills I relate apply to all leatherwork, and I've given directions for projects which use a variety of leathers.

Qualities of Leather

An understanding of how leatherworkers judge and compare leathers, and how they choose one for a particular use can be learned by examining four basic variables involved with tanning: *thickness, tannage, cut* and *type.*

THICKNESS

Leather can be had in a variety of thicknesses, from thin glove leather to thicknesses of an inch or more. The thickness is a primary consideration in choosing a leather, for most articles are most successfully made of leather of a particular weight. Thickness can go a long way toward determining flexibility and durability, although a thin leather can be made stiffer than a thicker one through certain tanning and finishing methods.

Weights of different leathers

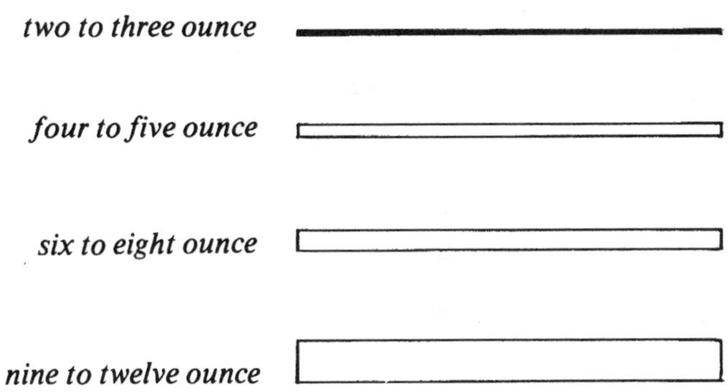

two to three ounce

four to five ounce

six to eight ounce

nine to twelve ounce

Leather thickness is gauged with two scales. Of the two, "ounces" and "irons," the former is the most commonly used. Ounces are used to gauge both *thickness* and *weight*, terms that have come to be used almost interchangeably. One ounce equals $\frac{1}{64}$ of an inch of thickness, and a square foot of leather which is $\frac{1}{64}$ of an inch thick weighs one ounce; a ten-ounce leather — in theory — weighs ten ounces per square foot. The Thickness Chart above will show you actual thicknesses.

In industry, an entire hide is machine "split," or cut horizontally into two or more uniform layers of a desired thickness. In home tanning, this is not possible, so plan to obtain heavy leathers from large animals, and lighter leathers from smaller ones.

TANNING METHODS

The method by which a leather is tanned imparts various qualities, as do the tanning and finishing agents. Generally, leathers that are oil-tanned are soft, supple and stretchy. Mineral-tanned leathers are relatively waterproof, while having some stretch. Vegetable-tanned leathers are preferred by many leatherworks because more may be done with them than with other types: Only vegetable-tanned leathers, for example, can be successfully tooled. They can be molded or stretched when wet, and will retain the shape when dry. What is more, vegetable-tanned leathers are easy to work with, take dye and finish well, and can be finished to be quite waterproof.

Phyllis Hobson's directions may be followed to tan any animal skin. But all skins (as any other raw material) have characteristic qualities and limitations, regardless of treatment. Deer hide, for example, can't be tooled and is naturally thin. Cow hide, on the other hand, is typically eight to ten ounces (in the unsplit, home-tanned version), and would hardly make a satisfactory pair of gloves.

You may wish to tan a specific hide so that you can make a particular project, but it may be more feasible to tan whatever skin you have on hand and see what application the resultant leather has.

CUTS OF LEATHER

The quality of the leather may remain fairly constant in a small skin, but in a large one, such as cowhide, the thickness and density vary considerably throughout. Commercial hides are sold divided up in several ways (see illustration).

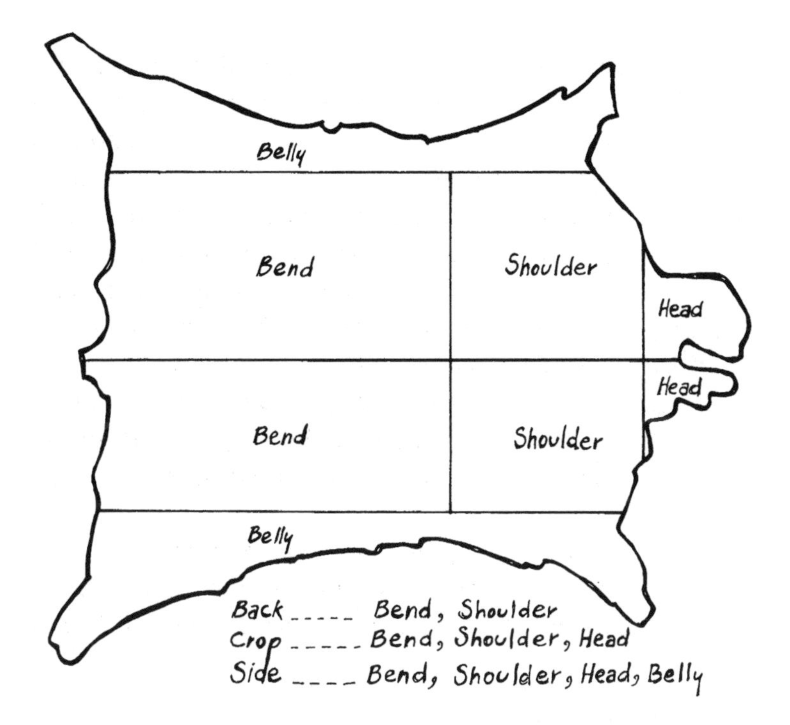

In industry, large hides are divided in half (down the backbone) early in the process, for easier handling. Each half is called a "side." Sides are often further divided into "bends," "crops," "shoulders," "backs," and "bellies," so the craftsman can buy leather of consistent character.

Bends are considered the finest leather in terms of evenness, texture, and durability. *Shoulders* come next, because, unlike the bend, wrinkles from the neck are often present. *Bellies* place last; they are "loose" and stretchy.

A *back* combines a bend and shoulder; a *crop* consists of a back plus the head; and a *side* is the crop plus the belly — one half of the entire hide.

You will probably tan a hide whole, or in sides. But the above information is important, and you should realize that different parts of larger skins have different qualities.

Small articles can be made from one skin of a small animal, or small skins may be pieced together to make larger articles, much as many small pieces of fabric go into a quilt. Numerous small articles, or a few larger ones, can be made from the skin of a large animal, for each side may measure 25 square feet or more.

TYPE

Venison tastes different from beef. Ever compare cow and goat milk? Leathers from different animals or animal groups have distinct characteristics, too.

Cowhide, from cows, bulls, calves and even unborn calves, is the type of leather used most often. It is uniform, durable, and very workable. It is also readily available (butcher your own cow or go to a slaughterhouse), and is used to make nearly every sort of leather article.

Equine hides, from horses, mules and donkeys, are similar to cowhides but are more durable and are coarser-grained.

Sheepskin is thin and strong. It is especially nice when tanned with the wool intact, and makes good garment leather.

Goatskin, too, makes good garments. It also is thin (two to three ounces) and very fine grained.

Pigskin, from domestic hogs, is usually considered to be of inferior quality, since it is pierced by characteristic hair follicle holes.

Deerskin, being soft and supple yet quite strong, has many garment uses. Having a simpler cell structure than many skins, it is easy to tan.

Hand Tools

Leatherworking is a craft with which you can involve yourself to any degree. You may make only a key fob or two, or you could end up raising rabbits and making coats from the fur skins. Either way, it is a satisfying craft, especially if you use leathers you tanned yourself.

Just as you have learned about tools that are necessary to tan leather, so, too, you will need specific tools to produce leather goods.

There are over a hundred different leather hand tools which you can purchase, and a complete set, of professional quality, may cost several hundred dollars. However, I would urge you to go slowly in developing your tool set. Start with a few tools, adding to them as needed.

In the following pages, I have listed, described, and illustrated several common leather hand tools which represent an investment of about 45 dollars.

This basic set is adequate to make most leather goods; this does not mean that other tools are superfluous, or that you won't eventually purchase them. Certainly, more sophisticated tools are useful and valuable in specialized or continued use.

BASIC TOOL SET

Utility Knife. An inexpensive purchase, this is the best common knife for all-purpose cutting. Good leverage can be applied and dull blades are easily resharpened or replaced.

Shears. Light leathers are most easily cut with good quality fabric shears.

Awl. This tool can be used to mark leather by scratching a line, or to mark holes to be punched with a prick mark.

Revolving Punch. Get one equipped with six "tubes." This punch lets you make six different size holes in one- to eight-ounce leather.

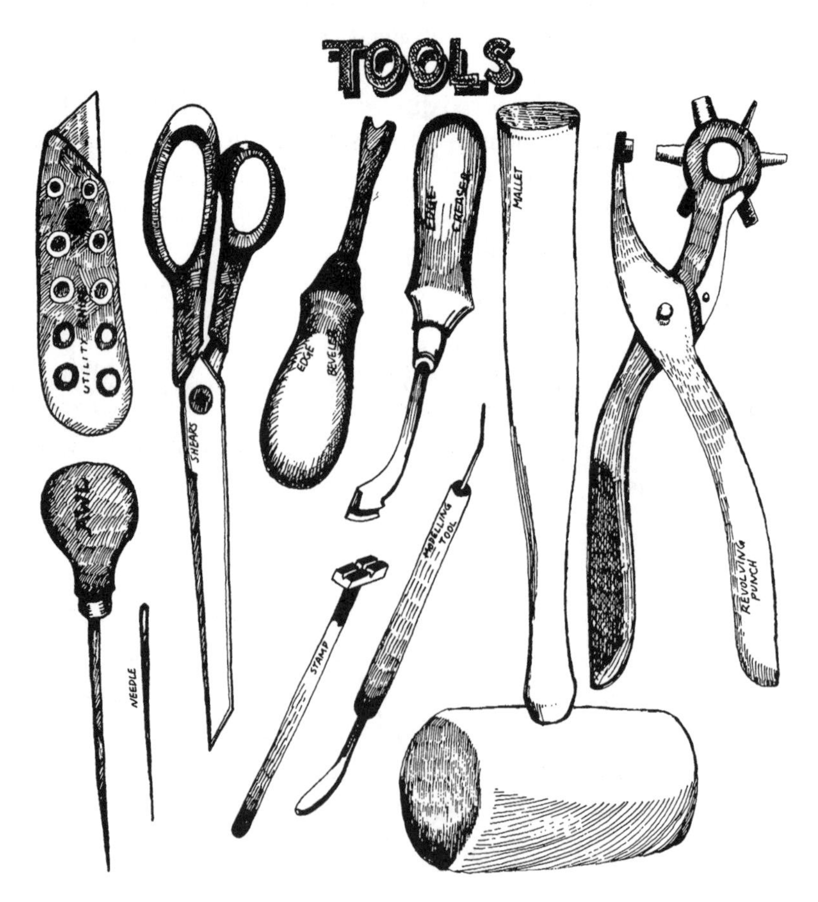

Edge Beveler. When leather is cut with a utility knife, the cut edge is square and looks unfinished. This tool rounds the corners, lending a more polished look.

Edge Creaser. This tool is used in conjunction with the beveler, and creates a decorative "bead" around edges.

Hammer. Any hammer will work for tooling leather or setting rivets. But a wooden or rawhide mallet won't mar the leather's surface.

Modelling Spoon. Modelling tools (see "tooling") come in many designs, but the spoon with a tracer end is the most basic. Buy others as needed, or make them of wood or rustless metal.

Stamps. Like modelling tools, stamps of many designs are used to tool leather (see "tooling").

Needle. There are several special leather needles, but any stout two-inch needle with a large eye will do.

SUPPLIES

Dye. Aniline dye is permanent and comes in many colors which may be blended. Leather may also be colored (with more or less success, depending on the leather) with fabric dye, ink, acrylic paint or permanent felt-tip markers.

Dauber. This is the simplest dye spreader around. Use one for each color dye.

White (Elmer's) Glue. This will hold leather well, but is not flexible. Use only where leather will remain stationary.

Barge All-Purpose Cement. This is a very strong, flexible, rubber-based contact cement. Available in quantities from ounces to gallons, it is excellent for bonding all leathers.

Thread. Waxed linen thread is preferred for medium and heavy-weight leathers, for it is very strong and the wax helps hold it in place. Carpet, or other heavy thread, may be used in lighter leathers.

Saddlesoap. Made of oil and wax, saddlesoap cleans, polishes and softens all finished leather.

Paste wax. A protective coating of wax should be built up (on finished leather only) with several buffed coats of quality paste wax.

FINDINGS

Rings. Ranging in size from ¼ inch to two inches, they are used as buckles or as a junction in a strap.

Buckles. These come in many widths and several styles and are used for fastening belts and straps. See drawing below, in which the single and double bar buckles require an oval slot to accept the buckle tongue.

Findings

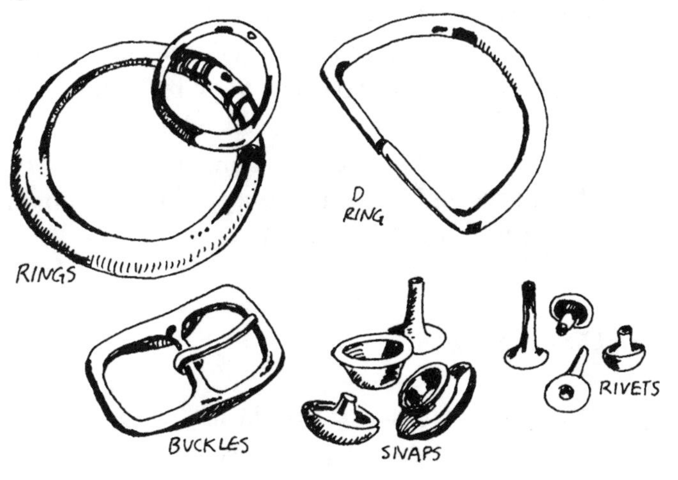

Snaps. Leather snaps are a decorative closing device. There are four parts which require a setter for installation.

Rivets. These permanent two-part fasteners are made up of a stud and a head. Set with a hammer, they come in several shank lengths and diameters, single or double-headed.

Findings are made from a variety of metals: solid brass, brass-plated, nickel-plated, copper, pewter or white metal. Solid metal findings are more expensive by far than plated ones, but they are usually a good buy since the finish cannot wear off, as always happens with plated findings.

Basic Handworking Techniques

Working with leather has many similarities to working with fabric, and anyone who sews should not have difficulty making the transition from cloth to leather. Others will soon catch on, as there are no exotic or complex *basic* techniques. (Twelve-part braiding comes later as an advanced technique.)

The following handworking techniques are used for both finished leathers and fur skins except where common sense dictates otherwise, or where noted.

PATTERN-MAKING

After a leather and the article to be made from it are chosen, make a full-size pattern of paper. Make a pattern piece for each piece of leather to be cut, even if pieces are identical.

Always work from a pattern, for the thought that goes into making the pattern can eliminate costly and frustrating errors after the leather has been cut.

There are patterns for ten leather articles in a later section of this book. Some are shown without scale and can be enlarged to any size you wish. Most are drawn to scale on grids, and may be enlarged to working size in one of two ways:

1. Buy or make one-inch-grid graph paper. Transfer the pattern, square by square, to size. This is somewhat painstaking, but can be very accurate, and actual dimensions are indicated to make it easier.

2. Obtain the use of an opaque projector, such as those used in many schools, churches and offices. Use this book and project an enlarged image on one-inch-grid graph paper attached to the wall. Measure the projected image to be sure it is adjusted to the proper size.

A commercial clothing pattern may be used, as is, for leather, particularly if the garment is not especially tight fitting. Some patterns need adjusting. Add to seam allowances, and add at least an inch to sleeves and legs, since leather or suede will eventually develop creases at elbows and knees, and will "draw up" more than the cloth the pattern was made for. Or, rip out the seams of an existing article and use the ironed pieces as a pattern. The article can be re-sewn afterward.

You may wish to make patterns for projects of your own design. Simple articles, such as watchbands, need only simple, one-piece patterns. More complex, three-dimensional articles, such as bags or garments, require more pattern pieces. Make a fabric or heavy paper dummy of your project before laying out your pattern on leather.

CHOICE OF LEATHER

As mentioned earlier, you may tan a certain leather to make a special project, or tan what you have on hand and see what you can make of it. In the latter case, common sense is a good guide. You will soon learn by feel what weight and softness leather will yield what projects. While certain articles require either very light leather (gloves) or very heavy (sandals), some articles can be made in nearly any weight.

PATTERN PLACEMENT

When your pattern pieces are all made, roll out your leather, grain side up. Inspect it for any objectionable spots, such as deep cuts, holes, discolored fur, spots, or thin places. Flip the skin over, flesh side up. Place the pattern pieces, avoiding any spots you noticed on the grain side. Work from the edges inward, rather than cutting a piece from the center of the leather. Arrange the pattern to make maximum use of the leather by butting up straight edges, and putting pieces very close together. Try several arrangements before you decide which is most practical.

Patterns may be arranged in any direction on medium- and heavy-weight leathers (4 to 12 ounces), but patterns should be laid "with the grain" on fur skins and on leathers of three ounces and lighter. The fur and grain run from what was the head to what was the tail, and thin leather stretches less this way than side to side. Cutting pieces in accordance with the grain will promote even, graceful draping in garments and light articles. When possible, lay your garment pattern so that the finished garment will fit you the way the skin did the animal it came from.

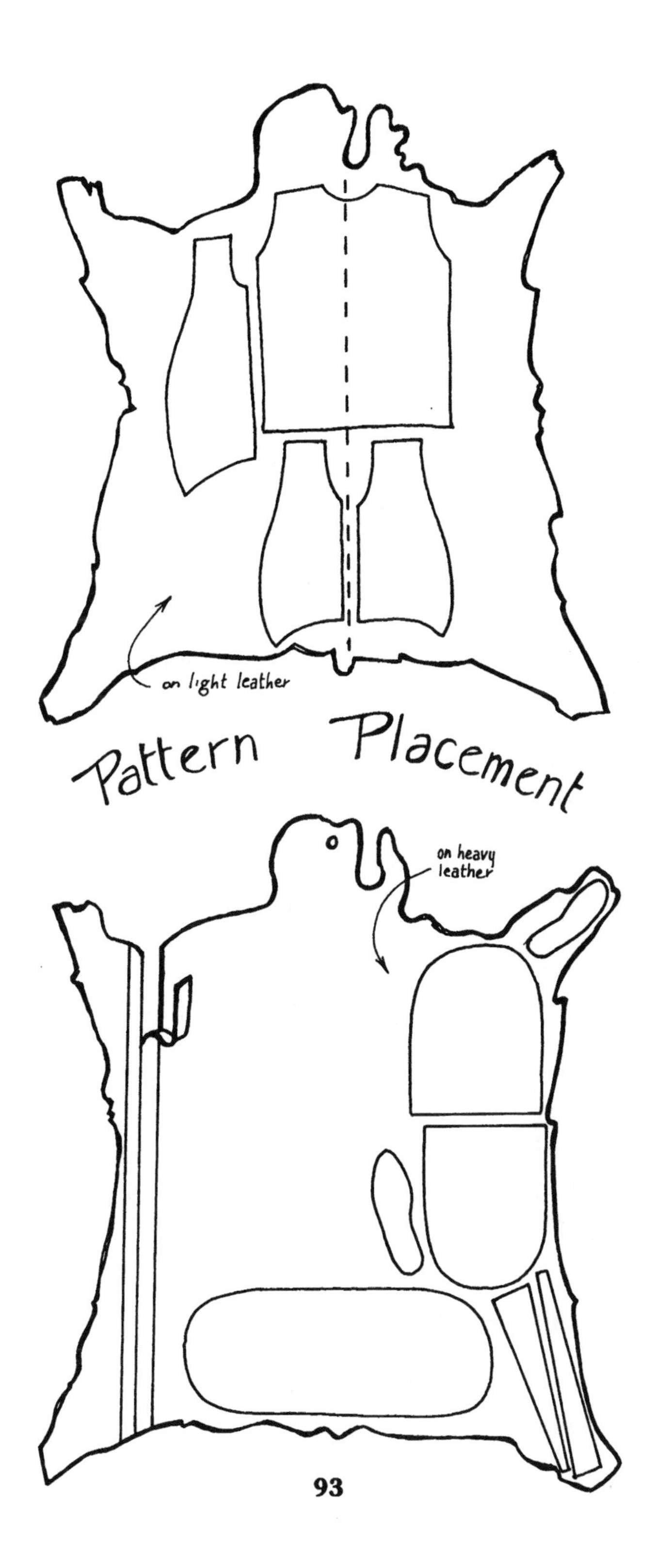
on light leather
Pattern Placement
on heavy leather

MARKING LEATHER

When you are satisfied that your pattern is arranged on the flesh side in the most sensible and economical manner, tape the pattern down, or adhere it with dots of all-purpose cement. Draw around each piece with a ballpoint pen. Felt tip pens are good, especially on fur skins, but check to be sure the ink doesn't bleed through. Mark holes for lacing or sewing by pricking the leather with an awl.

The patterns in this book indicate sewing holes on only one half of any given pattern. To mark the other half identically (which ensures hole alignment), fold the pattern and prick with an awl.

CUTTING OUT

Lay the marked leather on a large cutting surface, such as a half or whole sheet of plywood. Using your body weight, cut smoothly with the utility knife, making several passes. Don't cut into the leather which surrounds each pattern piece, for you may wish to use it for another purpose.

If the leather is quite light, you may wish to use shears, since a knife can pull or stretch a light leather, producing an uneven edge. If you are cutting a fur skin, the fur will hold the skin away from the cutting surface, making cutting difficult. Use a razor or very sharp blade to cut this skin, and be sure not to cut the fur.

HOLE PUNCHING

Holes are punched in medium to heavy leathers to accept thread, laces or findings. Make sure that holes align where they are supposed to, and that they are the right size. For thread, use the smallest tube in the revolving punch. Rivets call for a medium-sized one, and thongs take the largest.

Because of its short jaws, the revolving punch will punch holes only around edges. If you need a hole in the interior of a piece, see if cutting a slit will serve the purpose. If not, you will have to buy a "drive punch," a hole-punching tool which is hit with a hammer.

EDGING

Edging is a two-part method of giving your finished leather goods of four ounces and over a polished, more professional look. Well-finished edges, though a small touch, add much to the overall finished quality of a project.

The *edge beveler*, held at about 30 degrees, is pushed smoothly around the edges of the leather, paring off and rounding the knife-cut, 90-degree corner. It is followed in the same manner by the *edge creaser*, which creates a decorative crease or "bead" at the edge.

TOOLING

The appeal of working in leather for many, centers in the opportunity to *tool*, or make decorative (usually floral or geometric) bas-relief patterns in the surface of the leather (not fur skin) by modelling or stamping. To tool, you must use vegetable-tanned leather (four ounces or over), which will receive impressions when wet and retain them when dry.

Prepare the cut leather by "casing" it. Case leather by immersing it in cool water until it stops bubbling. Then wrap it, flat, in a towel and place it in a closed drawer. Depending on the thickness of the leather and the humidity in the air, the leather will be ready to tool some time within 24 hours.

Test it for readiness by pressing your thumbnail or a modelling tool into it. If an impression is left without raising any water, you may proceed. Since the heat of your hands will dry the leather as you tool it, keep an atomizer of clean water handy, and re-dampen as necessary.

While waiting for the leather to become ready to tool, prepare your tooling pattern on heavy tracing paper from other leather or pattern books or develop your own.

CUT, BEVEL, CREASE

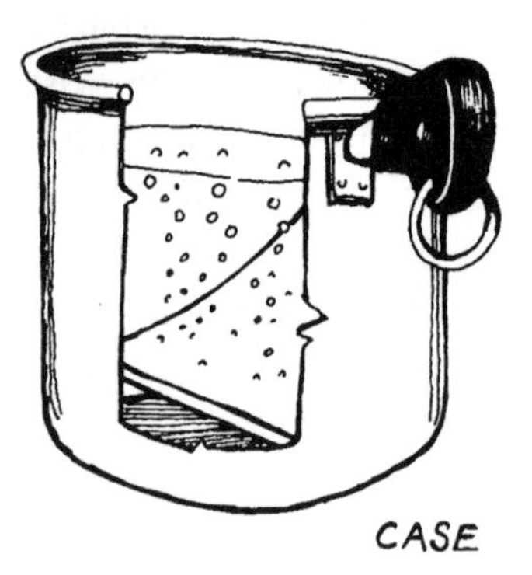

CASE

WRAP

TRACE

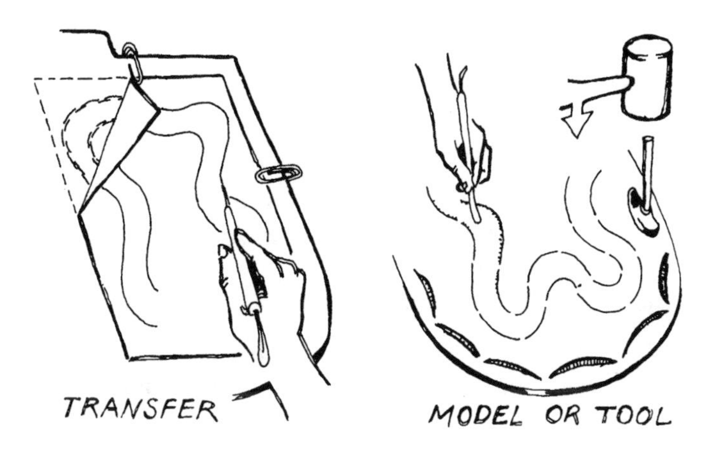

Place the leather grain side up on a hard surface. Lay the tooling pattern on top and hold it in place with paper clips. Impress the outlines with a dull instrument, such as a blunt pencil, a nut pick, or the tracer end of the modelling spoon.

When the outline is established, remove the pattern, and work directly on the leather, patiently deepening and refining the design, which is actually a bas-relief — a sculpture which gives the illusion of possessing much greater depth than actually exists, like the head on a coin.

Another style of tooling calls for the use of *stamps,* which you can buy or make of hardwood or nonferrous metal. Modular floral and geometric designs are built up with various stamps. The leather is placed on a hard surface — marble is ideal — and stamps are held in place and are lightly hammered, with the result that the impression of the stamp is left in the leather.

Many special techniques, such as *stippling* and *repoussé* are used in tooling, but space in this book does not permit a full exploration.

DYEING

In working with leather you have chosen a material that looks fine left natural. It also takes other colors exceedingly well. Furs may be dyed in the tanning, so this section is concerned with smooth leathers only.

Undyed leather typically is a light tan color that darkens with age, exposure to the sun, oils from the hands and other influences. Some people feel that the natural color (or at most, a light brown stain) is all the coloration any leather needs, and that its earthy beauty results in part from that simplicity.

My personal preference is for natural, undyed leathers. They look good at first, get nicer with age, and they don't fade or need touch-ups, as colored leathers do. Other people, perhaps of less rustic taste, favor their leathers in colors ranging through the spectrum. However you see it, leather takes color well.

Inks, fabric dye, leather dye, felt-tip markers, acrylic paints and craft paints all will color leather. So will water, coffee, tea, rust, blood and nearly anything else, unless the leather has a wax finish. Read the instructions that come with any coloring agent you wish to use, and make a test on scrap leather first, to avoid mistakes.

ASSEMBLY

The method of assembly, though one of the last processes performed, should be one of the first considered in planning an article to be made. Assembly, consisting of sewn, laced, or riveted seams, is an important element of design — as important as the form of the article or its color. Plan your assembly in your mind, on paper, in the pattern and in the leather—each type of assembly has its own characteristic effect. For example, fine linen thread looks much more finished than a rugged leather lace, although both can be correctly used in the proper context.

The illustration shows several machine- and hand-sewn thread and lace stitches and seams, and there are many others.

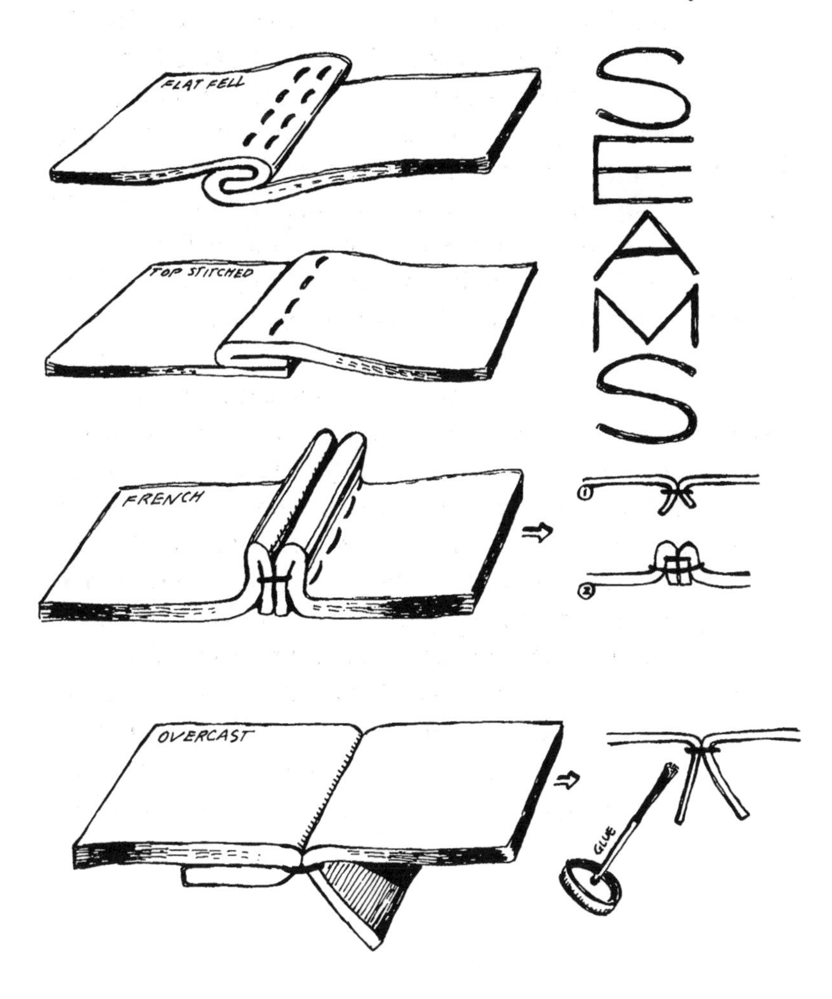

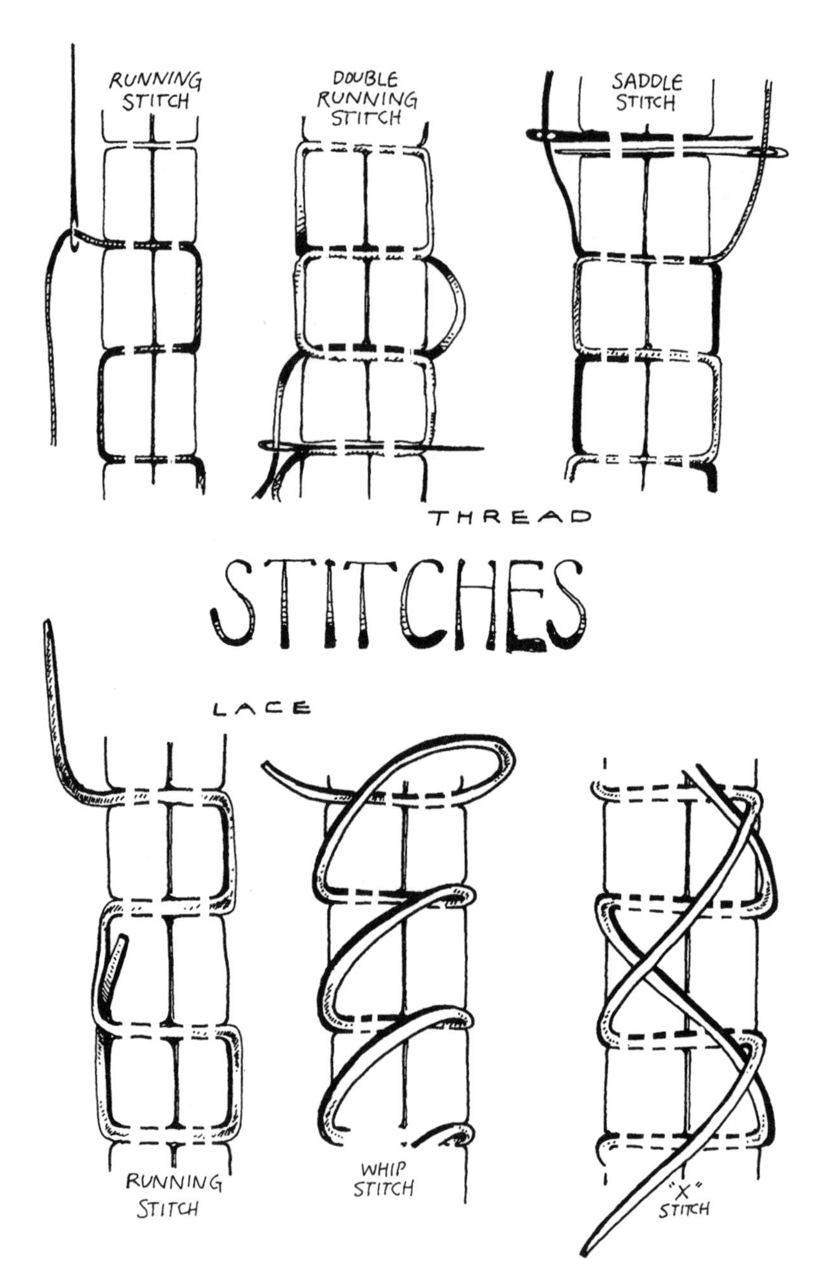
RUNNING
STITCH
DOUBLE
RUNNING
STITCH
SADDLE
STITCH
THREAD
STITCHES
LACE
RUNNING
STITCH
WHIP
STITCH
"X"
STITCH

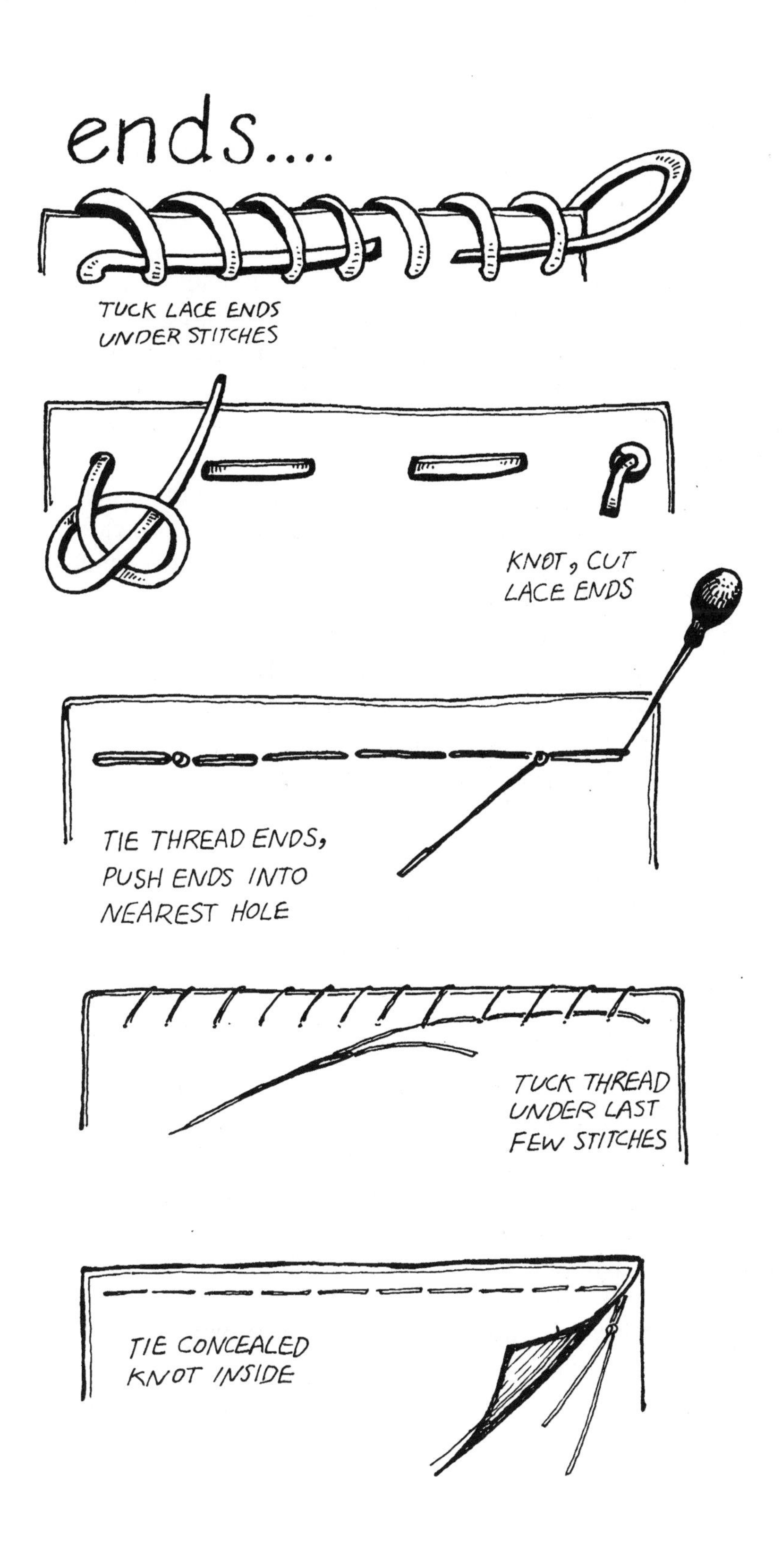
ends....
TUCK LACE ENDS
UNDER STITCHES
KNOT, CUT
LACE ENDS
TIE THREAD ENDS,
PUSH ENDS INTO
NEAREST HOLE
TUCK THREAD
UNDER LAST
FEW STITCHES
TIE CONCEALED
KNOT INSIDE

While each seam is performed differently, each should be sewn tightly, without bunching. Tie off thread ends with a square knot, and cut them three quarters of an inch from the knot. Push the thread end into a stitching hole, hiding it. "End off" lacing by allowing a two-inch end, and force it under three or four stitches.

Thin leather can be sewn on a good sewing machine, using a number 16 to number 19 leather needle, seven to nine stitches to the inch, with silk, cotton or linen thread. Baste the seams together first with all-purpose cement.

Some articles won't be sewn at all, but will be glued or glued and riveted. After punching rivet holes (if necessary), coat both surfaces to be bonded with all-purpose cement and allow them to dry. Be sure the alignment is right, since this is a contact cement and the pieces will adhere on touching.

FINISHING LEATHER

All leathers, no matter what tannage, weight, cut or type, are absorbent to some degree and will stain. To enhance and protect the beauty of your smooth leather (not suede or fur), a finish should be applied. Typically, a finish seals the surface, making it less porous, more waterproof. At the same time, a finish gives leather a soft shine . . . the look that made "the hand-rubbed finish" famous.

Mrs. Hobson has given several recipes for finishing and waterproofing leather. Others include beeswax, saddlesoap, shoe polish and paste wax. Most finishes are some combination of oil and wax.

As with coloring leather, there are many unknowns involved in finishing, so make a test on scrap of the same leather first. Normally, finishing is the last step (and should be repeated

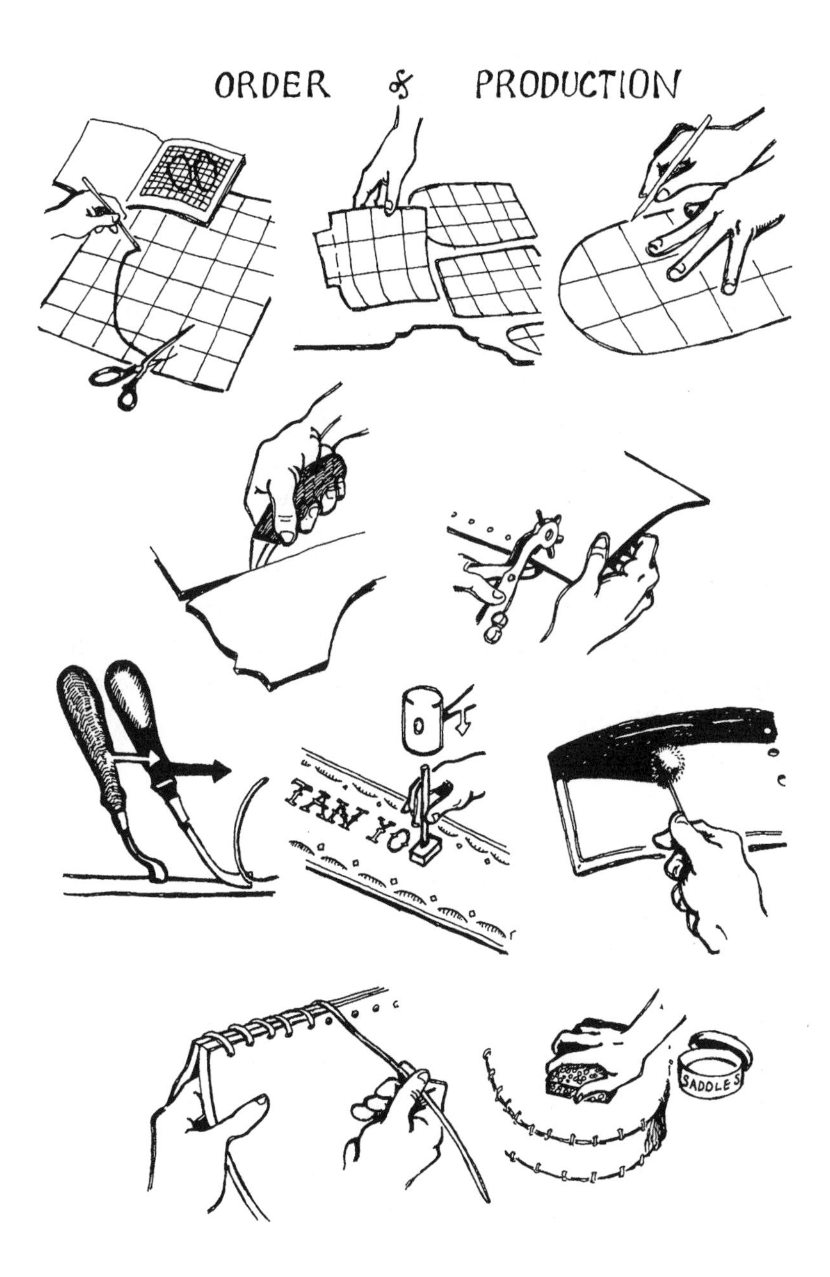
ORDER & PRODUCTION
TAN YO
SADDLES

yearly on all your leather goods), but there are times when a finish will be easiest to apply *before* assembly, when the pieces are still flat.

ORDER OF PRODUCTION

Generally, the basic handworking techniques I've given are performed in the order in which they were related: pattern making, pattern placement, marking leather, cutting out, hole punching, edging, tooling, dyeing, assembling, finishing.

But because all leather goods are different, it often happens that you will change the order or eliminate one or more steps. You may want to dye before punching holes if you object to having the dye run through and stain the back. When you tool, the edges come out nicer if you edge-bevel *before* tooling and edge crease *after*. (Edging does not apply to thin leathers and fur.) As mentioned, you may wish to finish before assembly. Remain flexible in your techniques and their order, and plan ahead!

Obtaining Tools & Supplies

Tan Your Hide is written for those who wish to tan their own leather and make things of it. This section is written for those who have tanned leather using Phyllis Hobson's directions, but it is not limited exclusively to home-tanned leather. Commercial leather may be used to make any article described here; the skills remain the same.

You must locate a leather retailer to buy leather. Many large cities have "leather districts": notably Boston, New York, Denver, San Francisco and Los Angeles. Most towns have a shop or two that retails leather as a sideline — a luggage shop, a large shoe repair or a handmade-leather-goods store. Be guided by the phone books found in your library.

You may buy in person or by mail. Write for a catalog, which is often very informative in itself. Order according to tannage, weight, cut and type.

Many leather retailers also sell tools, findings, and all sorts of supplies. Most of the tools and supplies listed in the basic tool set can be found at large department or hardware stores. Try a tack shop, farm supply or ship supply store for interesting and unusual findings.

PROJECTS AND PATTERNS

In this section I have given you actual patterns and directions for ten articles made of leather and fur. They range in complexity from basic to quite involved. From gun holsters to fur mittens, there should be something for everyone.

Leather Projects

1. KEY FOB

This is a good beginning project, for it is simple to make, while at the same time it uses many of the skills related in "basic handworking techniques." It is also functional and can be altered easily to suit your own design.

You will need:

a 2-inch by 3½-inch piece of 4- to 7-ounce vegetable-tanned leather

dye or acrylic paints

a split key ring (from a hobby shop)

rivets

Transfer the pattern from this book. Lay it on the leather, draw around it and cut it out.

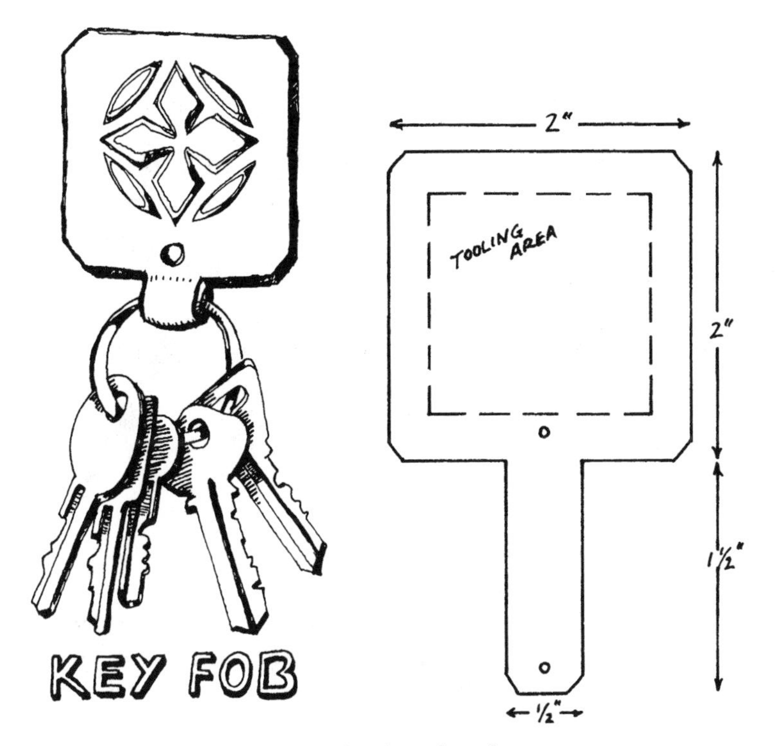

Punch the rivet holes. Edge bevel and crease.

Case the fob. While you are waiting, make a tooling pattern on heavy tracing paper.

Place the pattern on the leather and attach it with paper clips.

Proceed with the modeling, as in "tooling."

When the leather has dried completely, color it. Dye the whole fob a light color and work in dye of darker shades or other colors with a small brush. Or, mix acrylic paint with water to the consistency of heavy cream and apply it for brighter, opaque colors.

After the paint or dye has dried, fold the tab through the split key ring, set the rivet in the holes you punched, and hammer it home.

Finish the fob with paste wax.

2. GLASSES CASE

A case for reading or sun glasses is an easy, personalized project which is very handsome and functional when completed.

You will need:

one square foot of three- to five-ounce leather (vegetable-tanned if you intend to tool)

waxed linen thread or thin, commercial leather laces.

dye

saddlesoap or paste wax

Transfer the pattern from this book. Note the areas marked "Tooling Areas." These are the places to tool the pattern you choose.

Make a paper pattern and place it on the flesh side of the leather.

Mark the outline and the holes. Cut it out evenly.

If you intend to use waxed linen thread, punch the holes with the smallest tube in the revolving punch. If you decide to use laces, choose an appropriate tube.

Edge bevel and crease the leather.

Case it, make a tooling pattern, and tool the glasses case as in "Tooling."

If you dye the leather, you can emphasize the tooled pattern by painting the lowered areas with darker dyes than the higher, surrounding areas. Use a small brush and blend dyes on the leather by lightly scrubbing with a small stiff brush — a toothbrush works well.

Fold the glasses case so that the seam matches. With waxed linen thread, sew any of the stitches shown on pages 99-101.

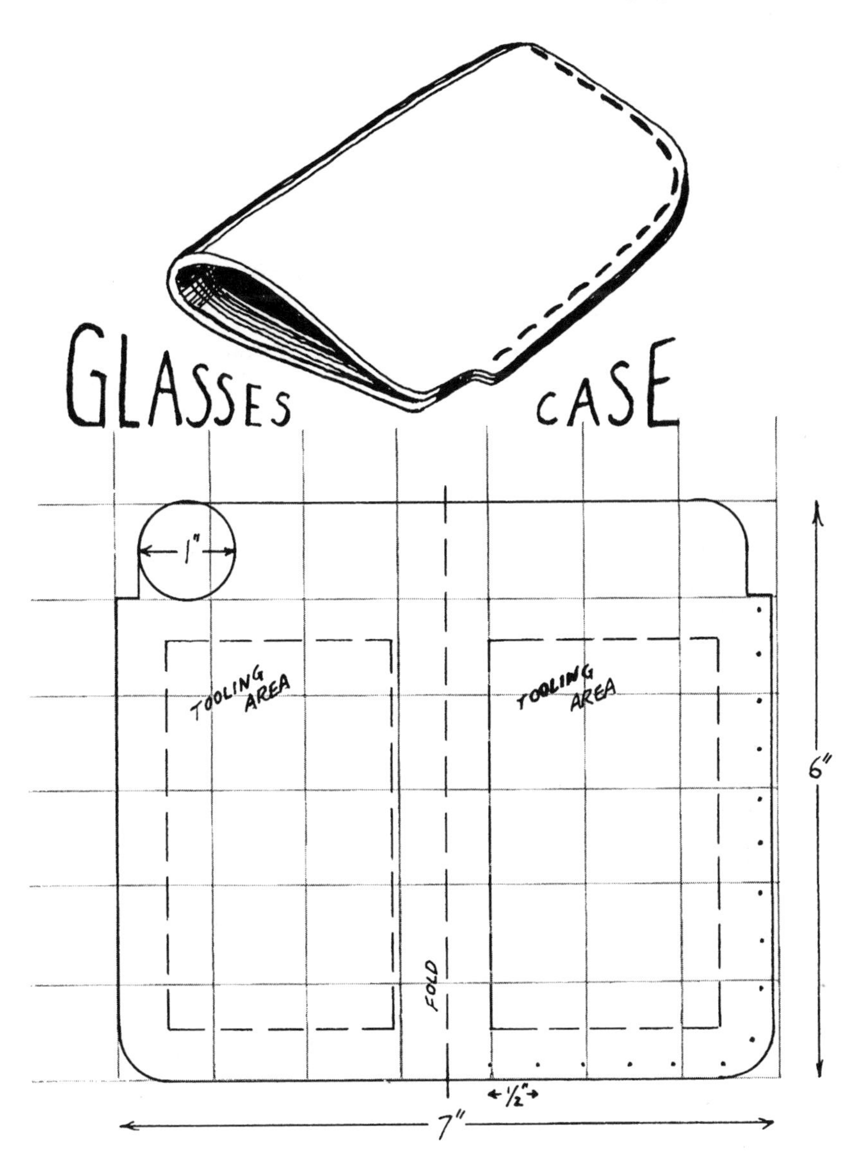

To lace, choose a lacing stitch from page 100. Leave three-quarters to one inch of extra lace at each end of the seam. Don't tie the ends off, but feed them back under several stitches to hold them tight.

Finish with saddlesoap or paste wax.

3. KNIFE SHEATH

Make a sheath for your skinning knife from a hide you removed with that knife. This particular pattern is designed for an eight-inch hunting knife with a four-inch blade. It is easily adjusted to fit larger or smaller knives. The "fringe insert" is optional and may be eliminated without changing the pattern at all.

You will need:

one square foot of medium (four- to five-ounce) leather

rivets

waxed linen thread

dye (optional)

saddlesoap or paste wax

Transfer the pattern from this book, and make a paper pattern. Lay it on the flesh side of your leather.

Mark it, including the holes, and cut it out. If you are using the fringe insert, cut it as a solid piece, then cut slits to establish the fringe.

Edge bevel and crease the leather.

Punch the sewing holes with the revolving punch's smallest tube, and the rivet holes with a medium tube.

Dye the leather, or leave it a natural color.

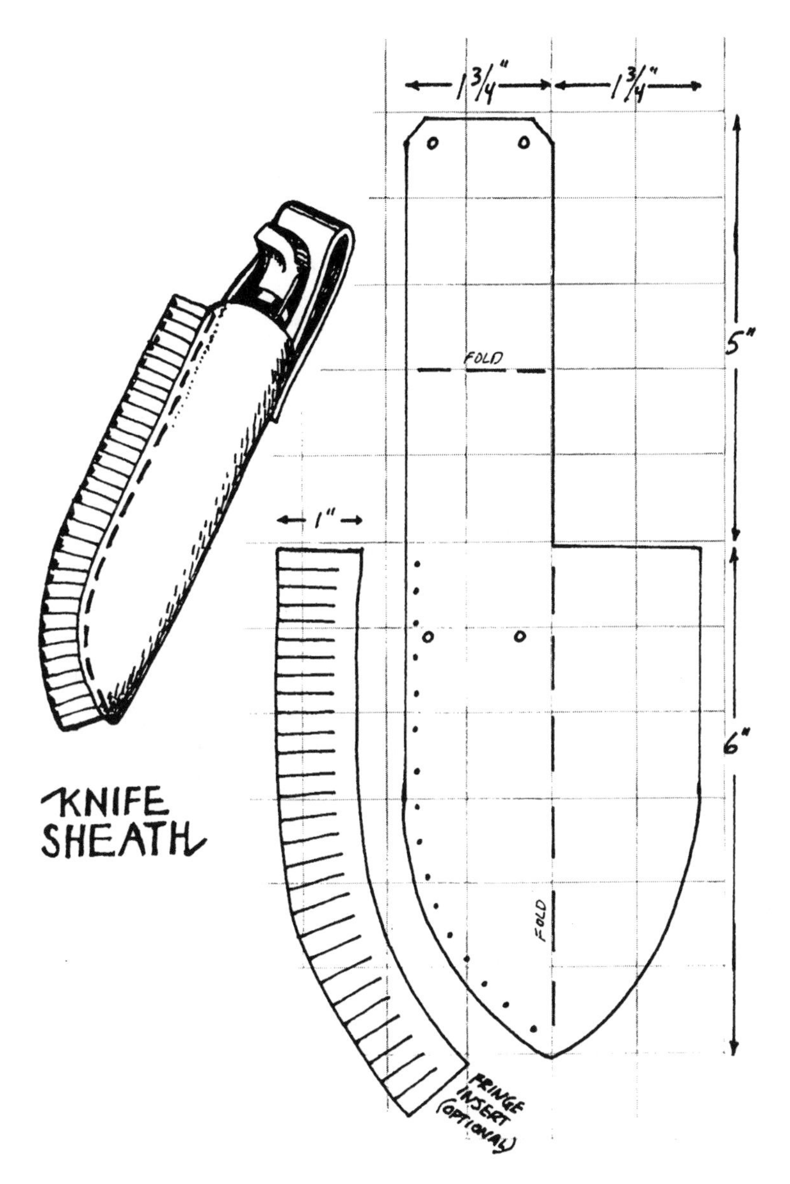

Fold and rivet the belt loop.

Fold the sheath so that the seam lines up. Sandwich the fringe insert between the front and back, and sew a double running stitch. Tie and tuck in the thread ends.

Rivet the seam ends, with the rivet caps toward the front.

Finish with paste wax or saddlesoap.

You can easily add a leg strap. Cut a 24-inch long, ¼-inch-wide thong. Punch a large hole in the tip of the sheath. Run the thong through. Weave it into itself, as illustrated below.

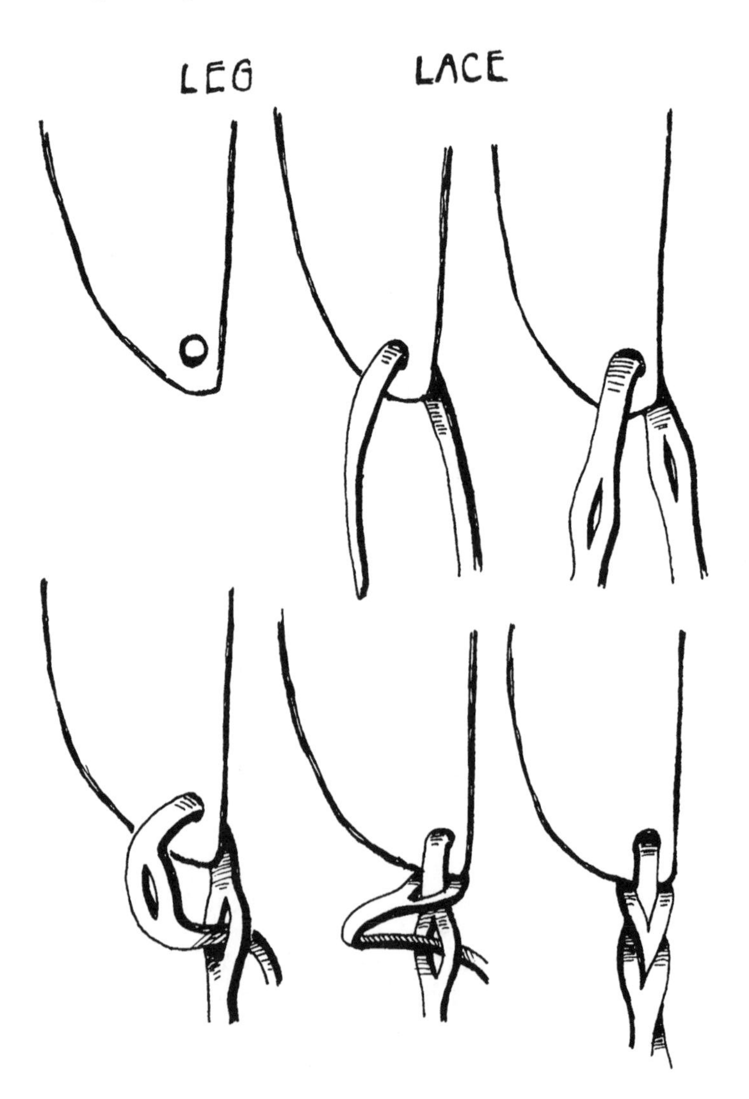

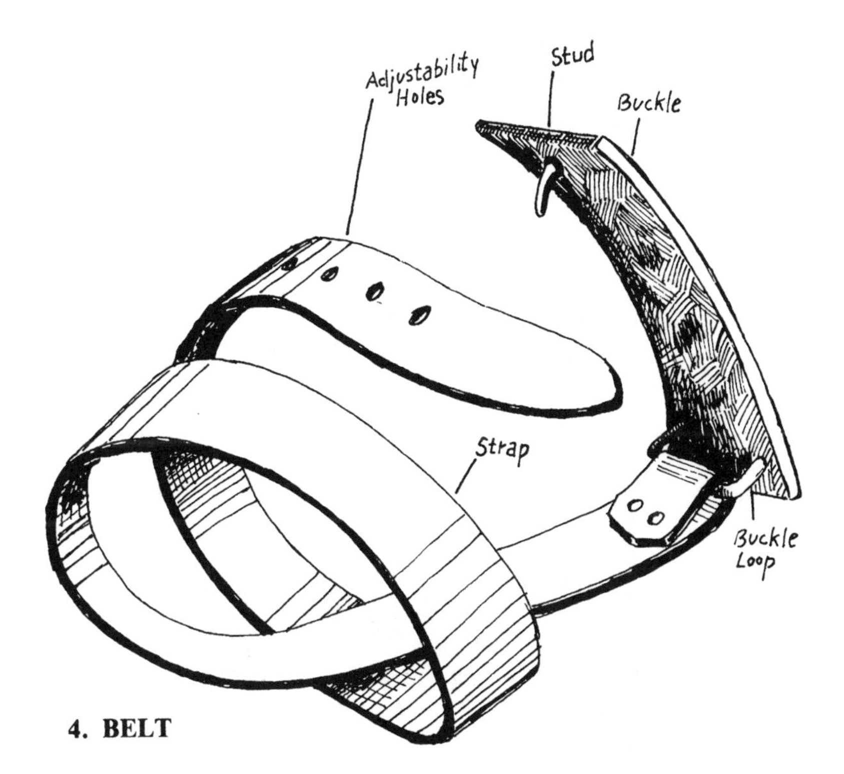

4. BELT

Within the rather rigid design and structural guidelines for a belt, there is much variation possible in the application of tooling, dye, patterns of carved leather or rivets.

To make a belt, use:

heavy (six- to eight-ounce) leather with little stretch

thread or rivets

a buckle

Buy or make your buckle first. Of three basic sorts, I'm dealing here with a stud tongue buckle. If you wish to use another type of buckle, the basic difference is that the others require an oval slot to accommodate the buckle tongue.

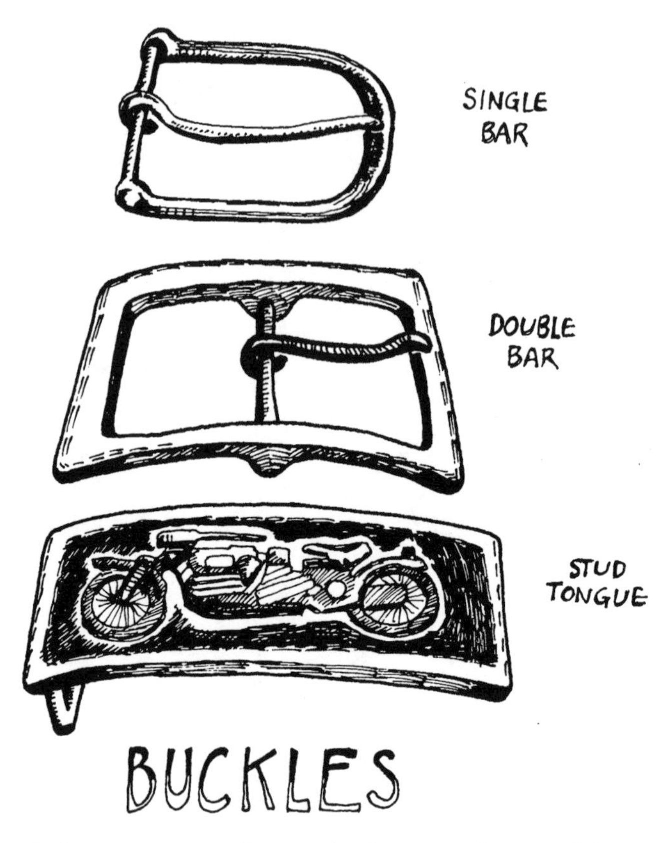

Since belts are narrow but quite long, and are subject to strain, the leather you cut from should be firm and stretchless. While short pieces may be spliced to form a longer strap, a shoulder will yield somewhere between one dozen and two dozen belt straps of the proper lengths, depending on their width and the actual measurement of the shoulder.

Above all else, belt straps must be straight. Establish a straight edge on your leather with a yardstick or a long, straight board. Draw and cut along it.

Straps have many uses in leatherworking, and you may decide to devote one skin solely to their production, since strap after strap can be cut from the established straight edge. Or, cut straps from one edge and pattern pieces from the opposite one.

Measure the interior width of the loop on your buckle. Standard sizes are in quarter-inch increments from one to two inches. Mark this measurement (minus ⅛ inch) in several places along the straight edge of the leather.

Connect the marks with your wooden or metal straightedge. and carefully cut out a perfectly straight strap.

Add seven inches to your waist measurement to find the actual strap length. This allows two inches to turn through the belt loop, and five inches at the other end for adjustability holes.

Cut the strap to the proper length. Trim both ends to give a more finished look. Simply cut the corners from the buckle end. The loose end may be cut at an angle, rounded or tapered.

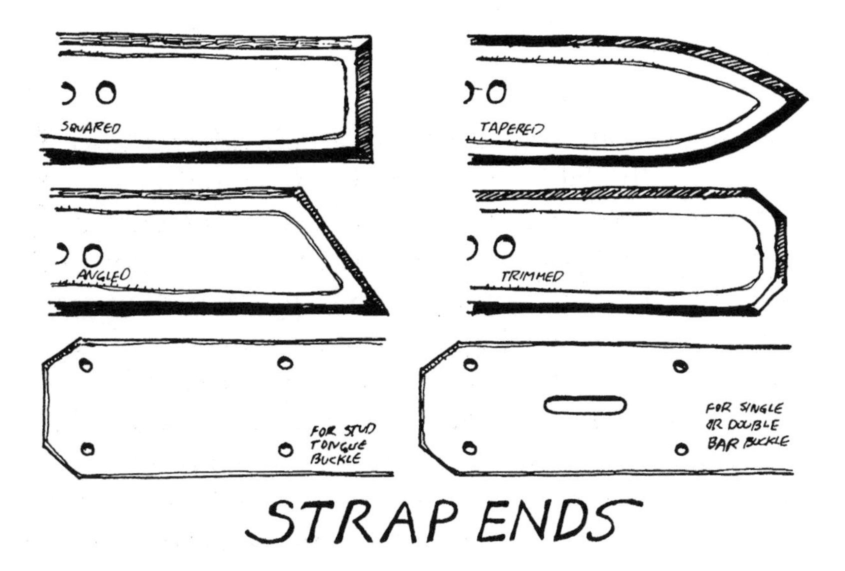

STRAP ENDS

Punch medium-sized holes for rivets to hold the buckle. Make two evenly spaced holes in the end first, then run it through the buckle loop. Fold it over and mark through the already punched holes with an awl. Punch them out.

Edge bevel and crease.

Tool the belt in sections, or with one pattern running the length. Or decorate with rivets or studs. Or leave it plain.

To dye the belt, dye the face first, then stand it on its side and dye the edges a darker color to accent them.

Mount the buckle, and rivet it down carefully to avoid cutting or marring the leather.

Finish with saddlesoap or paste wax.

Put the belt on, and hold it so that it is tight enough to do its intended task. Press the buckle stud into the leather behind it.

Remove the belt, center that mark within the width of the belt, and punch it out using a tube large enough to accommodate the stud. This hole becomes the center one of five adjustability holes. Measure so that each is one inch apart, and punch them as well.

Finish with paste wax.

5. WALLET

There are many sorts of wallets for many different tastes. They vary in basic design and in the materials that go into them. Because they are all small and functional and are carried on the person, they will require lightweight leathers and careful workmanship.

To make this wallet, use:

one square foot of light (two- to three-ounce) leather

Wallet

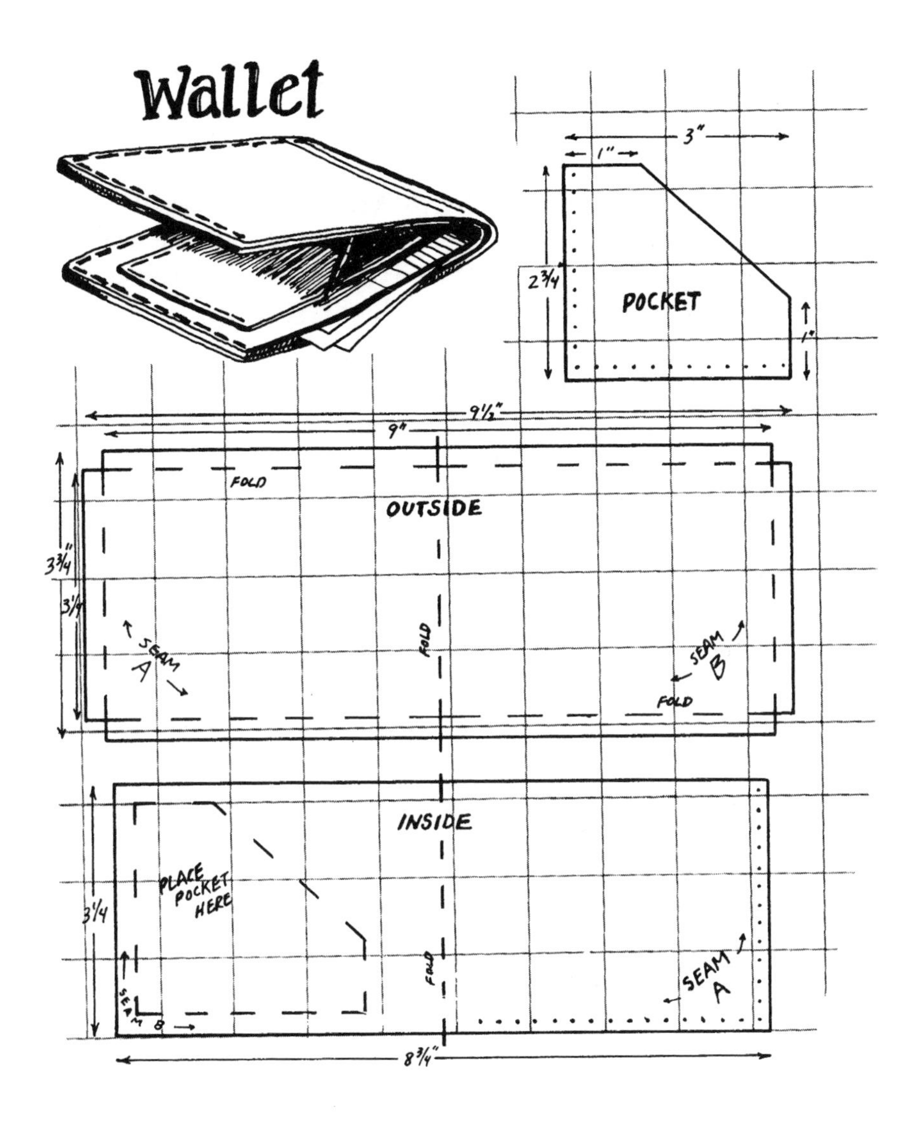

extra heavy-duty carpet thread

dye

paste wax

Some thin leathers are quite soft and these aren't suitable for making wallets. Choose a leather that has some degree of toughness and is firm, while retaining some flexibility.

Transfer the pattern. Make patterns for both pockets from the one pattern shown. Note that the "inside" is ¼-inch *smaller* than the "outside," necessary so that the wallet will fold without bunching. Because the pieces are different sizes, the holes for sewing will appear not to align.

Lay out the pattern on the flesh side of the leather in the most economical and practical manner.

Cut the pieces with a utility knife or shears.

Dye, or leave natural.

Fold the tabs on the back inward and glue them.

Punch out all the holes for sewing.

Sew the pockets on the inside, using a double running or other appropriate stitch.

Place the inside on the outside, flesh sides together.

Sew seam A with the same stitch used on the pockets. Tie and tuck the thread ends.

Align seam B. Because the inside is shorter than the outside, you will be forcing the wallet to fold into its characteristic shape. Tie and tuck the ends.

Moisten the leather fibers with a liberal application of saddlesoap, and create a fold by placing a weight on the wallet, overnight if necessary.

6. GUN HOLSTER

For those readers who not only "make their own" and "tan their own," but also "shoot their own," included here is a pattern for a handgun holster designed for a .44 Magnum "Super Blackhawk" by Sturm, Ruger and Co.

To adapt it to another gun, first make a heavy paper pattern. Tape it together, and see if your gun fits. If the holster is too

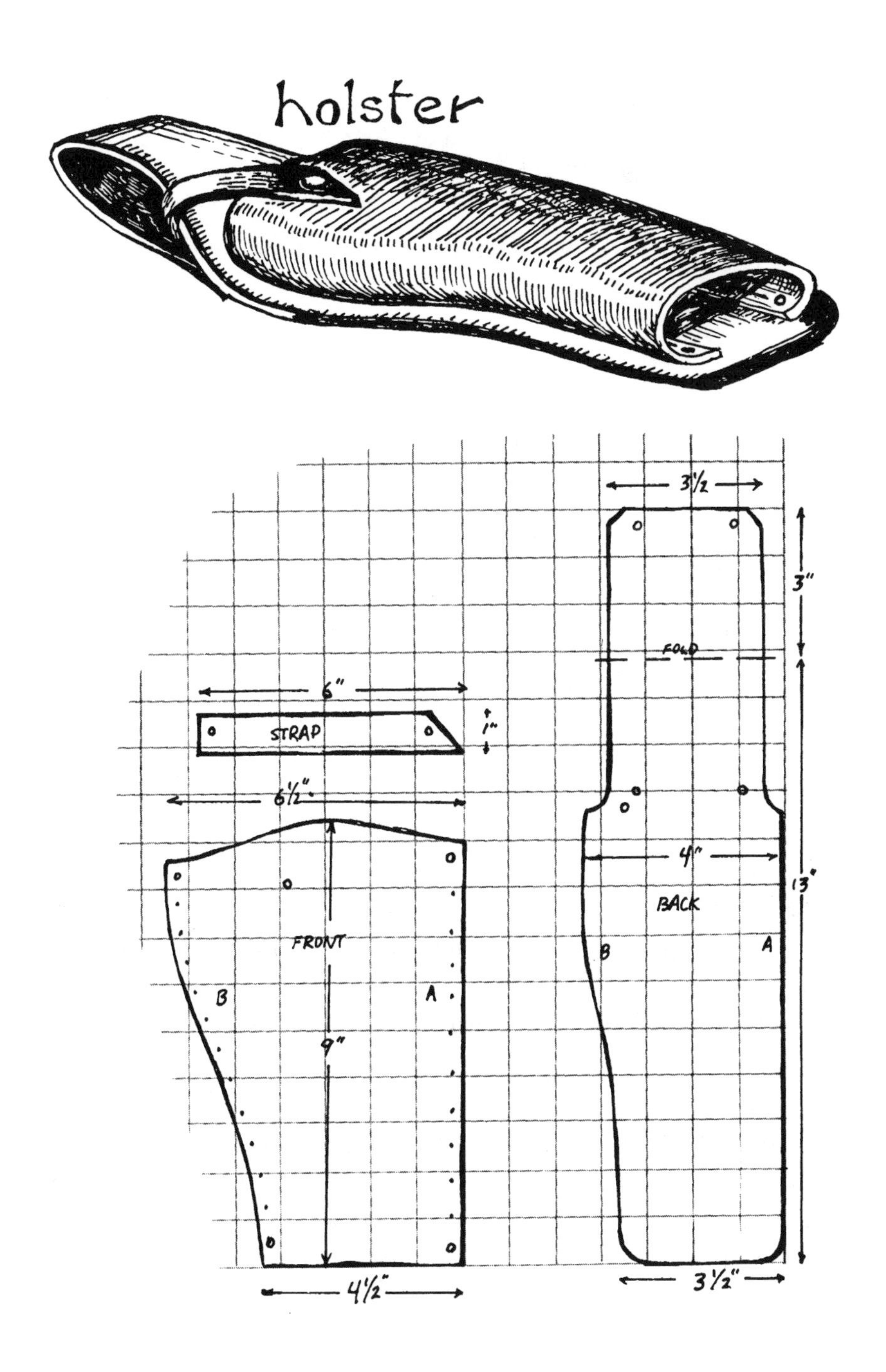
holster
6"
STRAP
1"
6½"
FRONT
B
A
9"
4½"
3½
3"
FOLD
4"
BACK
B
A
13"
3½"

large or too small, it is a simple matter to mark it accordingly on the paper mock-up and make a new pattern.

To make this holster, you will need:

two square feet of medium to heavy leather: four- to seven-ounce

rivets

snaps

waxed linen or other heavy thread

Establish your pattern and cut it out of paper. Make a dummy holster, and when you have adjusted it to size, lay it out on the flesh side of your leather.

Mark the pattern outlines with a pen and the holes with an awl.

Cut out the leather and punch the holes.

If your revolving punch won't reach the hole for the snap in the "front," make a small slit with your utility knife.

Edge bevel and crease all the pieces.

If you wish to tool your initials or a pattern in the holster front, proceed as described in "tooling."

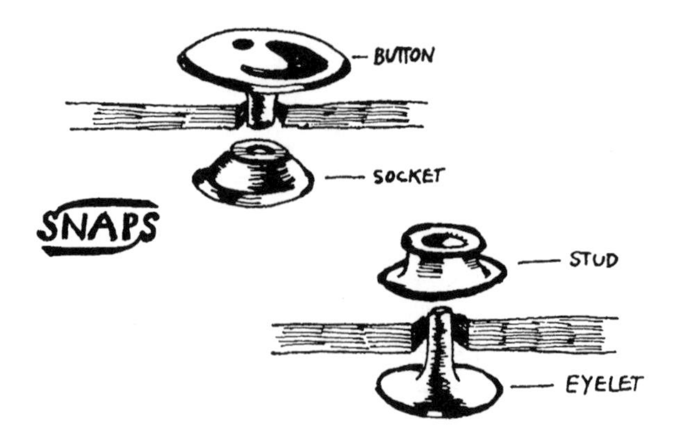

Dye the leather, or leave it a natural color.

Begin assembly by first installing the snap on the front and the strap. Put in your own, or a shoe repairman will install one for you.

Fold the belt loop over. Insert and hammer rivets, with the caps facing the back.

Rivet the strap to the back, adjusting the length to your gun.

Prepare your needle and thread; use a little more than four times the length of the seam.

Note that the front of the holster is turned under at the edges, so neither seam A or B will show. Place the front on the back, grain sides together.

Line up seam A and sew a double running stitch. Rivet the last hole on each end.

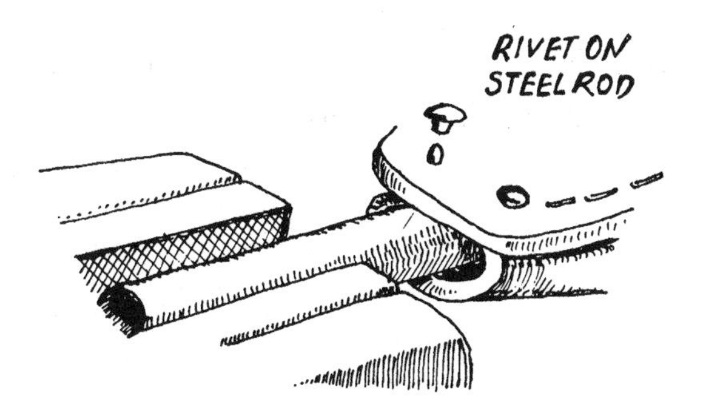

"Turn" the front, so that seam B lines up. Begin sewing at the narrow end. Since the space is small, you will find it easiest to sew the entire seam very loosely, then draw the stitches tight with a bent wire.

Insert rivets in the end holes. To set them, use a steel rod held in a vise as an anvil (see illustration on page 121).

Finish the holster with paste wax.

Fur Skin Projects

1. SHEEPSKIN PILLOW

Looking quite elegant and appearing more difficult to make than it actually is, this wooly pillow is a comfortable addition to any room. It can be made any dimension — from sofa to floor size.

You will need:

sheepskin: 2½ square feet for a 12- by 12-inch pillow, or
9 square feet for a 24- by 24-inch pillow, or
20 square feet for a 36- by 36-inch pillow

a pillow form of the appropriate size

extra heavy-duty carpet thread

Estimate the size and buy a pillow form first.

Make a paper pattern. I've included a pattern here, but it is so direct that you can make your own by measurement, rather than by grid transfer. Be sure all the corners are square, and allow at least a four-inch overlap.

Sheepskin Pillow

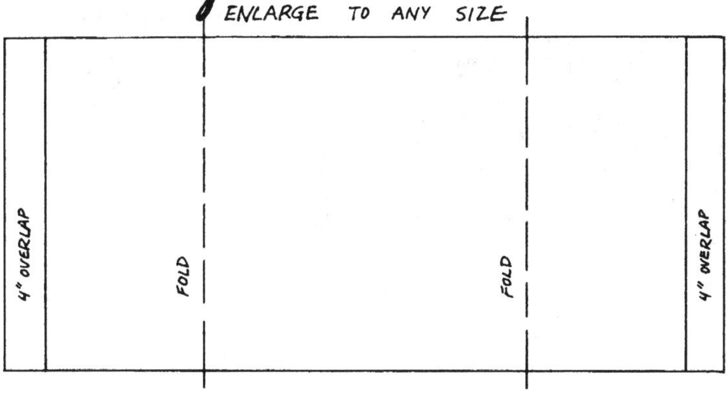

ENLARGE TO ANY SIZE

4" OVERLAP
FOLD
FOLD
4" OVERLAP

Lay out the pattern on the flesh side of the sheepskin. If it is too large to fit on one skin, piece together at the fold lines, adding a half-inch seam allowance to each side to be joined.

Mark the pattern, including the fold lines, on the sheepskin.

Cut carefully, and avoid cutting the wool.

Fold on the fold lines, flesh side out. With extra heavy-duty carpet thread, sew a running stitch, six to the inch.

Turn the pillow right side out, and insert the pillow form. The overlap needs no closure, for the tautness of the form keeps it closed. The form can easily be removed and the sheepskin can be cleaned when necessary.

2. FUR HAT

This fur hat is a warm one which can be made of almost any skin tanned with the fur intact. Designs of this sort are typically executed in sheepskin, which, because of the length of the hair, may be the warmest type of fur and the most comfortable to wear.

Because this hat is rather loose-fitting and can be turned up, one size fits many heads. Make a muslin dummy to be sure it

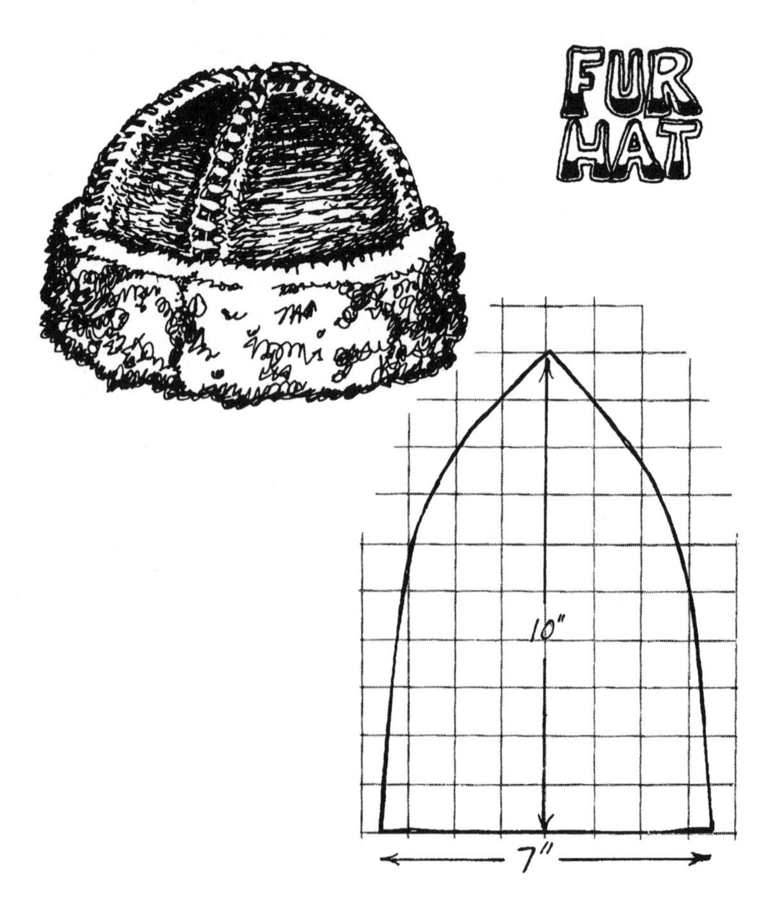

fits. If necessary, you can make the hat larger or smaller by increasing or decreasing the seam allowances.

You will need:

four square feet of fur skin (two- to three-ounce)

extra heavy-duty carpet thread

Make your pattern of paper: all four pieces.

Lay the pattern pieces on the flesh side of the skin, and transfer with a ball point or felt tip pen.

Cut the pieces carefully, from the flesh side, to avoid cutting the fur.

This hat is designed to have fur inside, with flesh side and the seams outside. It is warmest that way. But it may also be reversed (fur out). If you can't decide which way you want it, sew it with the fur inside, and the seams (emphasized with a whip or X-stitch), outside. When you tire of that, just turn it inside out and have a hat with fur out, seams in.

Place two pieces together, seam to seam. Trim long fur close to the skin up to one-quarter inch from the edge. Hold edges together for sewing by basting with all-purpose cement. Or use paper clips (or even straight pins since the holes they make won't show).

Sew a whip or X-stitch, four to six to the inch, using thread of a complementary color. Reinforce the seam ends by sewing the last stitch double.

Tie off the ends, turn up the edge and wear.

3. FUR MITTENS

These fur mittens rival any others for warmth and charm, and are designed to match the fur hat on pages 124-25. They can be made of the same fur skin and thread, or of suede with removable wool mitten liners.

Like the hat, they are warmest and most comfortable when sewn fur in, seams out, but can also be sewn reversed.

The pattern given is for a "medium" hand, and it can be made larger or smaller at the seam allowances. No stitching holes are marked, since thin leather can be sewn by hand without punching holes first.

You will need:

six square feet of two- to three-ounce fur skin

extra heavy-duty carpet thread

Only one hand pattern (three pieces) is given. Be sure to make a right hand paper pattern *and* a left hand paper pattern. Simply reverse the pattern given to make the other hand pattern.

Mark all six pieces on the fur's flesh side.

Cut carefully, and don't cut the fur.

Trim long fur within a quarter-inch of all seam allowances.

Sew one mitten at a time. Baste the seam of the "thumb" and "palm" of one hand together, flesh sides out.

Sew a whip stitch or X-stitch, four to six to the inch.

Baste the seam of the palm and back together, flesh sides out, and sew the same stitch as used on the thumb.

Repeat the same procedure with the other mitten.

Tie and secure all thread ends. Turn the cuff up, exposing the fur interior.

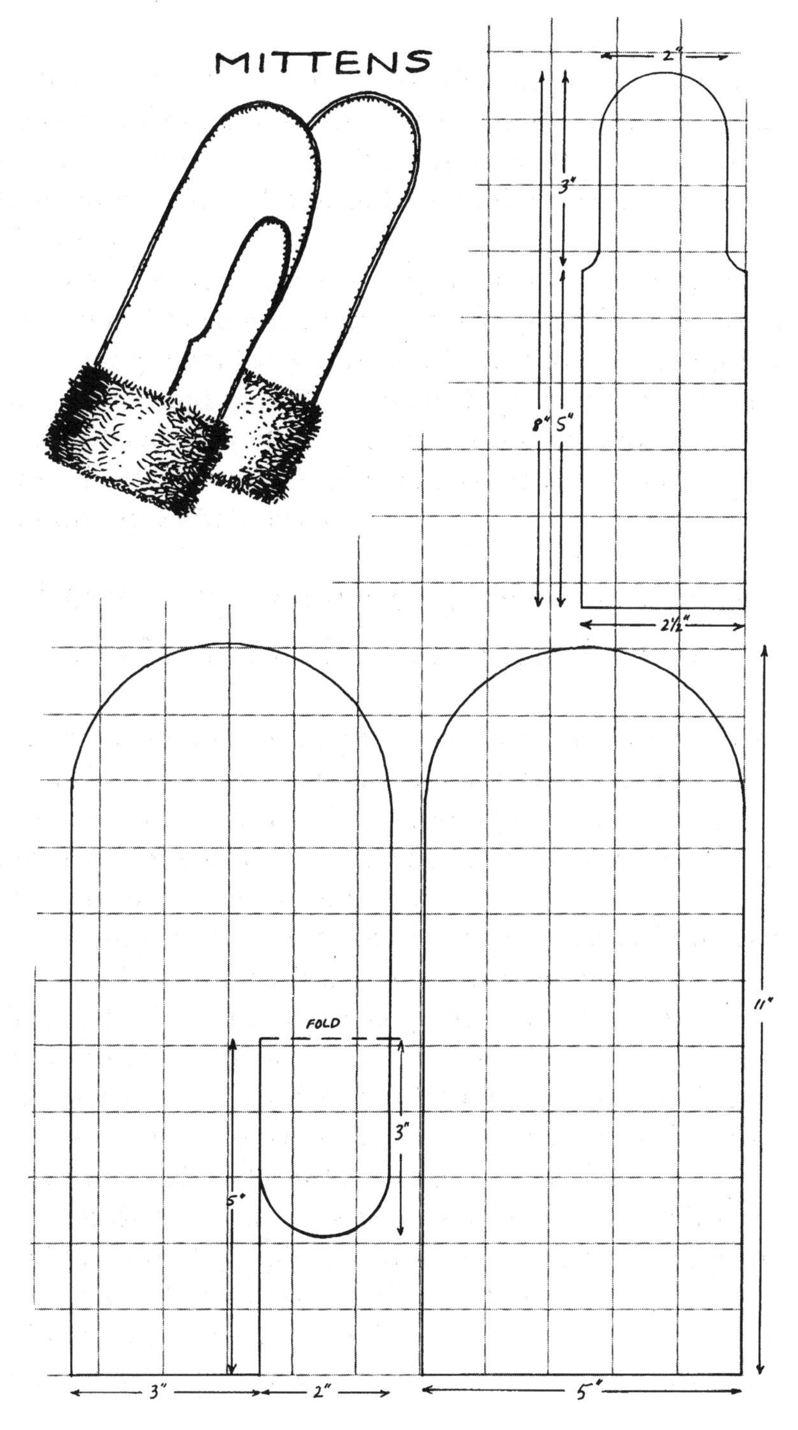
MITTENS
2"
3"
8"
5"
2½"
FOLD
3"
5"
11"
3"
2"
5"

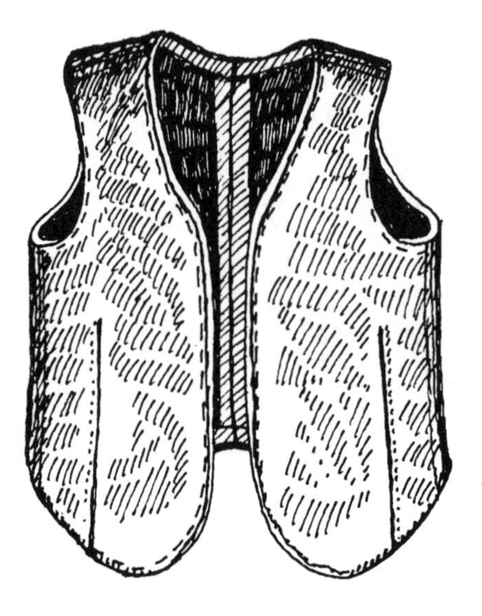

4. LEATHER OR FUR VEST

A leather or fur vest is a project which may be executed in a wide variety of soft, lightweight skins, with or without fur — sheep, deer, goat, cow, pig, rabbit, squirrel or other.

To make this vest you will need:

eight square feet of two- to three-ounce leather or fur

extra heavy-duty carpet thread, or silk, cotton or linen thread used with a number 16 to 19 leather machine needle

The pattern shown is for a man's vest in a medium (thirty-eight inch chest) size. A garment such as this must fit well, so by all means, make and adjust a dummy vest of inexpensive muslin. Or make it of lining material and sew it in later as a vest lining. Adjust the size at the seam allowances.

You can also arrive at a workable pattern by making a pattern from an existing vest. Take it apart (reassemble later)

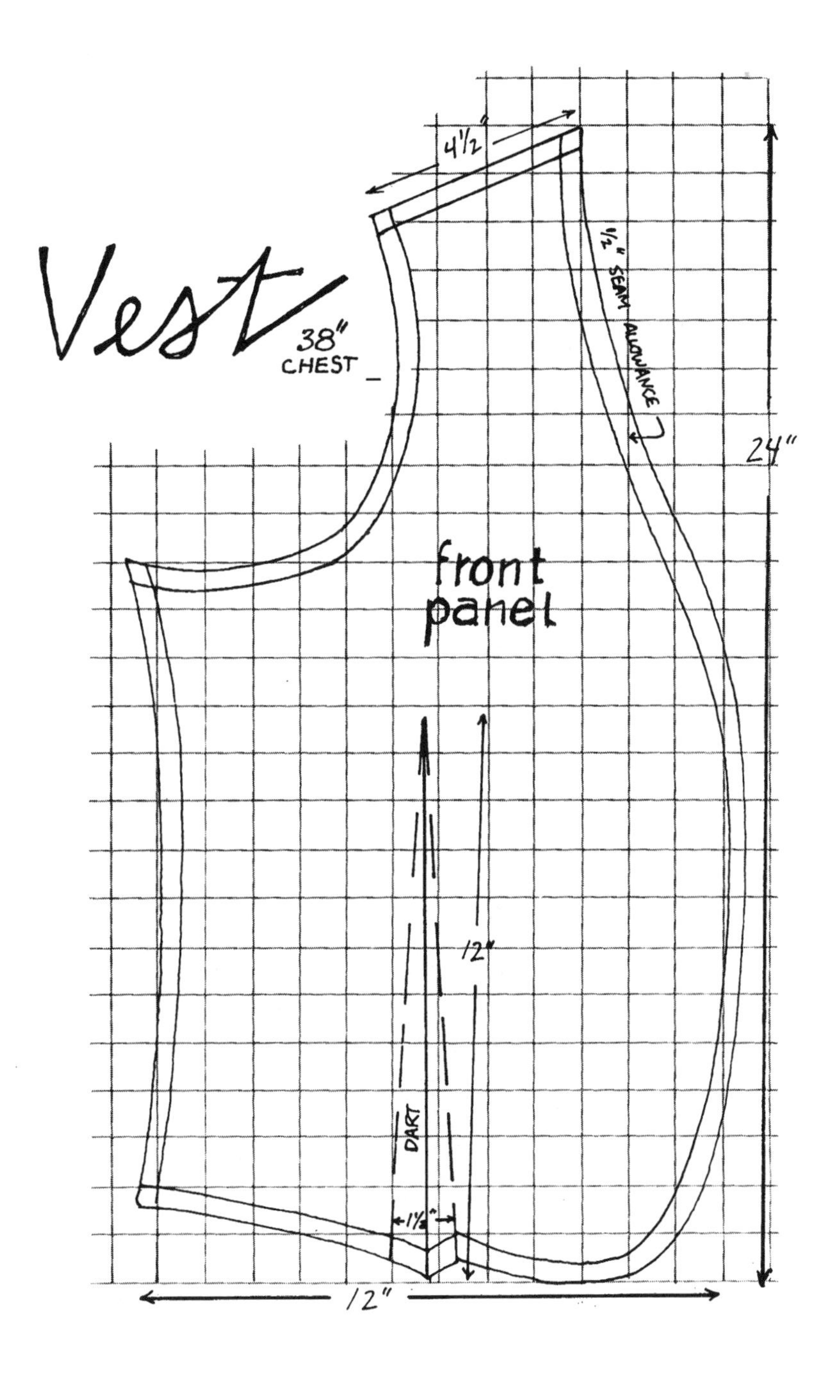
Vest
38"
CHEST
4½"
½" SEAM ALLOWANCE
24"
front
panel
12"
DART
1½"
12"

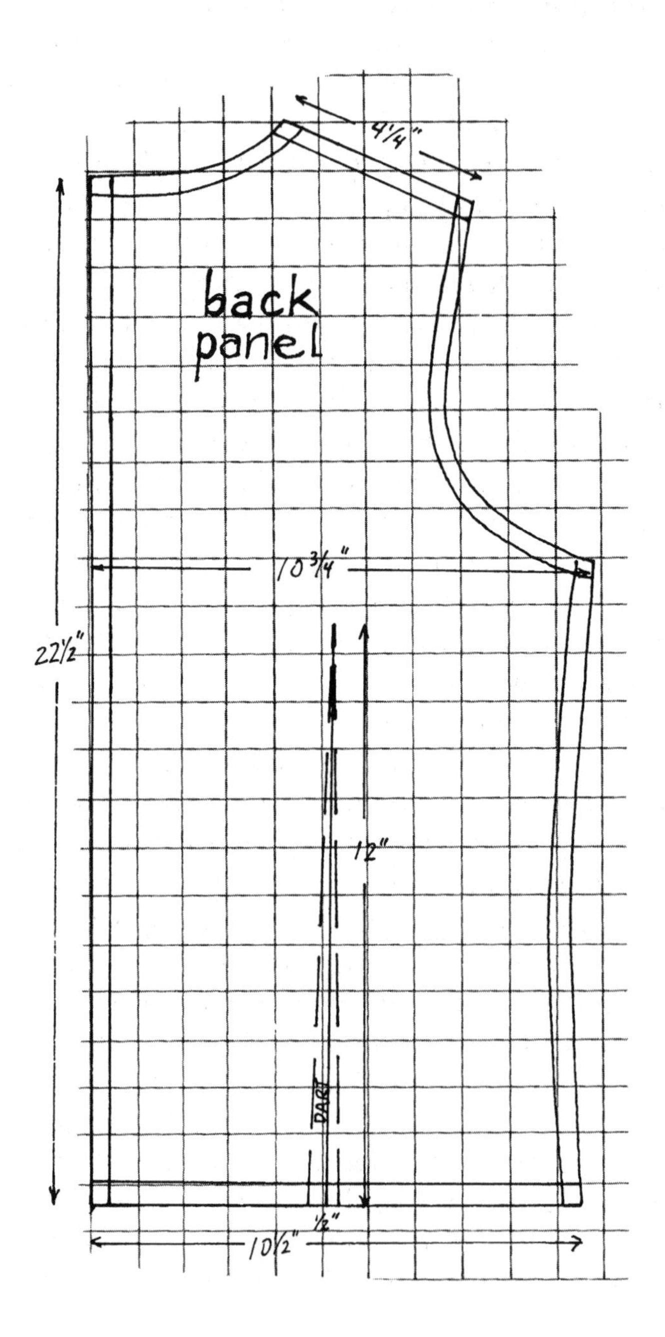
4¼"
back
panel
10¾"
22½"
12"
DART
½"
10½"

and use the pieces for pattern pieces. Or lay the vest on paper and draw around each pattern piece; add one-half inch all around for seam allowances. Vest patterns are also sold commercially at fabric and department stores where you can buy a pattern and style that fits you.

Make a paper pattern, and lay it on the flesh side of the leather. I have shown only two pattern pieces — "front" and "back" panels. Make paper patterns for both right and left front and back panels — four pieces in all.

Remember that it is desirable to have the grain run in the garment the same as it did on the animal, to eliminate problems with the grain of the fur, uneven draping or stretching. If possible, position the vest "back" over and in line with the backbone line on the skin, close to the head. Lay the vest's "left" and "right" panels on either side of or below the "back," running in the same direction. (See illustration on page 93.

Mark the leather and cut it out.

The most satisfactory for joining garment pieces is an overcast seam (see page 99), which may be sewn by hand or machine. After sewing the seams, open the leather and glue the seam allowances back. Sew the darts on the indicated lines; then cut the darts open and glue the leather back.

Turn the edge seam allowances and glue them down. Topstitch a decorative running stitch, or other stitch. Punch holes first if the leather is too thick to comfortably sew by hand.

Tie and tuck in all thread ends.

Smooth leather (but not fur or suede) should be given an application of saddlesoap to soften and preserve the leather.

INDEX

Other Storey Titles You Will Enjoy

The Backyard Homestead Guide to Raising Farm Animals,
edited by Gail Damerow.
Expert advice on raising healthy, happy, productive farm animals.
360 pages. Paper. ISBN 978-1-60342-969-6.

The Backyard Lumberjack, by Frank Philbrick & Stephen Philbrick.
Practical instruction and first-hand advice on the thrill
of felling, bucking, splitting, and stacking wood.
176 pages. Paper. ISBN 978-1-58017-634-7.

Basic Butchering of Livestock & Game, by John J. Mettler Jr., DVM.
Clear, concise information for people who wish to slaughter their
own meat for beef, veal, pork, lamb, poultry, rabbit, and venison.
208 pages. Paper. ISBN 978-0-88266-391-3.

The Beginner's Guide to Hunting Deer for Food, by Jackson Landers.
Everything a first-time hunter needs to know, from choosing
the right rifle to field dressing and butchering.
176 pages. Paper. ISBN 978-1-60342-728-9.

Storey's Basic Country Skills, by John and Martha Storey.
A treasure chest of information on building, gardening,
animal raising, and homesteading — perfect for
anyone who wants to become more self-reliant.
576 pages. Paper. ISBN 978-1-58017-202-8.

Storey's Illustrated Breed Guide to Sheep, Goats,
Cattle, and Pigs, by Carol Ekarius.
A comprehensive, colorful, and captivating in-depth guide
to North America's common and heritage breeds.
320 pages. Paper. ISBN 978-1-60342-036-5.
Hardcover with jacket. ISBN 978-1-60342-037-2.

Wild Turkeys, by John J. Mettler Jr., DVM.
Complete habitat and mating information for
hunting or observing these noble birds.
176 pages. Paper. ISBN 978-1-58017-069-7.